Abel Hernández-Muñoz

Zoology, Animal Science

Abel Hernández-Muñoz

Zoology, Animal Science

An overview of this science within everyone's reach

ScienciaScripts

Imprint

Cover image: www.ingimage.com

This book is a translation from the original published under ISBN 978-613-9-43440-4.

Publisher:
Sciencia Scripts
is a trademark of
Dodo Books Indian Ocean Ltd. and OmniScriptum S.R.L publishing group

120 High Road, East Finchley, London, N2 9ED, United Kingdom
Str. Armeneasca 28/1, office 1, Chisinau MD-2012, Republic of Moldova, Europe
Printed at: see last page
ISBN: 978-620-8-04791-7

ZOOLOGY, ANIMAL SCIENCE

MSc. Abel Hernández Muñoz

Table of Contents

INTRODUCTION

The biosphere is the terrestrial sphere in which all living components of the Earth are found in close interaction with the atmosphere, the hydrosphere and the lithosphere. This sphere of life is characterized by the great diversity of organisms in correspondence with the way they adapt to their living conditions, as well as their unity in relation to the common characteristics that exist among them.

Each of the living components of the Earth have adapted to particular conditions in which the typical structures of each species respond to the habitats they occupy. Relationships are established among all of them, as well as with the non-living components of the place they inhabit, which makes ecological equilibrium possible, provided that the preservation of environmental conditions so permits.

Since the emergence of life on the planet, and with it man as the most evolved organism in the zoological scale, the study of different organisms has been a constant concern of scientists. Spontaneous empirical knowledge allowed the human species to discover characteristics and behaviors of the different organisms in nature and to be able to act in correspondence with that knowledge. But nothing could replace the need to discover the essential characteristics and reactions of organisms among themselves and with their environment. With evolutionary development, new species appeared and some resembled others in form and structure. In response to the needs of man himself, some dedicated themselves to the study of the development of life on the planet and thus biology emerged as a science with its system of laws, categories, principles and relationships with other sciences that identify it as such.

"One of the first aims of biology was to establish broad generalizations about living organisms, so that useful knowledge could be passed on from generation to generation. Early in the history of man it was useful to know which animals were dangerous, which were good as food, which produced or came to transmit disease, and so on. Soon it was observed that living organisms had certain peculiar characteristics by means of which they could be easily identified and grouped into differential categories."[1]

Zoology is the science that studies the behavior of animals and their interrelationship with the environment, in close relationship with each other. It also studies the evolution of species and the transmission of hereditary characters from one generation to another. He has also had to classify the different species in different categories with the purpose of grouping them in correspondence with their most common characteristics. It is also dedicated to the study of their diversity in correspondence with the different types of ecosystems.

Accordingly, Zoology has different branches that correspond to the analysis of all the factors inherent to the different zoological groups. In essence, it studies the different groups of animals in terms of their anatomy and physiological functions.

Today we find ourselves in the 21st century, a time of momentous changes in biological diversity brought about by man-made disasters. Over the last few decades, research in the basic science of biology has generated astonishing knowledge about ourselves, the human species, and the millions of other life forms we share on the planet. The application of this basic research has led to the technology for organ transplants, genetic engineering, the eradication of numerous diseases and increased global food production. Recent studies in molecular biology and genetics have produced new knowledge about diseases. This century has seen the importance of biology in improving the quality of life enjoyed by many human beings.

As zoologists continue to study the interrelationships of the animals that inhabit the planet, humankind's awareness of their effects on other organisms and their environment expands.

This book is structured in chapters that go from zoological diversity to the more general study of the relationships of organisms in their environment.

---

[1] Valentín Arbona, Marta. Systematic Botany I. Editorial Pueblo y Educación, Havana City, 1988.

Self-monitoring questions

1. What does Zoology study?
2. What are the main branches of Zoology?
3. What is the importance of Zoology for the scientific development achieved by mankind?

CHAPTER 1: THE UNITY AND DIVERSITY OF THE LIVING WORLD

For all those who are introduced in the study of Biology and specifically Zoology, it is always interesting and a concern, to explain why there is so much diversity in nature as far as animals are concerned. To be able to offer an accurate answer, it is first necessary to understand how life arises on the planet; but it would be impossible to understand this without first analyzing aspects related to the primitive Earth.

Origin of life.

It is the opinion of many that life originated only once, and in environmental conditions very different from those that exist today. Various researchers have offered their results through hypotheses. One of the most widely accepted is the one on which Oparin, a Russian scientist, based the possible origin of life.

It is believed that at first the Earth was cold, but as gravitational compaction continued, heat accumulated, to which energy from the radioactive decay of some elements contributed. This heat sometimes escaped through hot springs and volcanoes, which also produced gases. These constituted the second atmosphere of the early Earth. It was a reducing atmosphere, with little or no free oxygen. These gases included carbon dioxide (CO2), water vapor (H2O), carbon monoxide (CO), hydrogen (H2), and nitrogen (N2). It is also possible that the early atmosphere contained some ammonia (NH2), hydrogen sulfide (H2S), and methane (CH4), although these reduced molecules may have been rapidly degraded by ultraviolet radiation from the Sun. As the Earth slowly cooled, water vapor condensed until torrential rains began to fall, forming the oceans. The precipitation eroded the Earth's surface, adding minerals to the oceans and causing them to become salty.

Four requirements for the chemical evolution of life were met:

-The absence of free oxygen.

-Energy.

-Chemical elements.

-Time.

Oxygen is very reactive and would have degraded the organic molecules that are a necessary step for the origin of life. However, because the Earth's atmosphere was strongly reducing, any free oxygen would have formed oxides with other elements.

A second requirement was a high-energy location, with violent thunderstorms, widespread volcanism, meteor bombardment, and intense radiation, including high ultraviolet radiation from the Sun. It is thought that the "young" Sun produced more ultraviolet radiation than the present Sun, plus the Earth lacked a protective ozone layer to block much of that radiation.

Third, the chemicals necessary as constituents for chemical evolution must have been present. These included water, dissolved inorganic minerals (present as ions), and the gases of the early atmosphere.

A final requirement was sufficient time for the molecules to accumulate and react.

The age of the Earth, about 4.6 billion years, is a suitable time for chemical evolution that occurred in several phases. First, small organic molecules formed spontaneously and accumulated over time. They were able to accumulate instead of being degraded (as they are today) because the two factors that now degrade organic molecules-free oxygen and other living things-were absent on the early Earth. Second, large macromolecules, such as proteins and nucleic acids, were assembled from smaller molecules. Then, the macromolecules interacted with each other and assembled into more complex structures that over time were able to metabolize and duplicate. Later, these macromolecular assemblies became cell-like structures that ultimately became the first true cells.

In Oparin's theory, it has been considered that the upward and downward movements of the water masses of the primitive seas (tides) left deposits in the hollows of the relief near the coast; at times of low tide, the external factors already analyzed above were responsible for triggering the processes already described and consequently the origin of very simple organic molecules called prebiotics, which over time became increasingly complex, the amino acids, that is, the constituent elements of proteins, and the nucleotides that form nucleic acids. Over time these molecules continued their evolution and with it their variations.

In their complex evolutionary process, the first manifestations of life must have been characterized by being covered by a selective membrane that guaranteed the exchange with the environment and the presence of a genetic code favoring reproduction in exact replicas of the progenitors. At the beginning, where not all these characteristics were present, what were called progenotes, that is, very primitive structures similar to a cell, were not considered true cells. The very process of adaptation to diverse environmental conditions and the appearance of DNA as part of the cellular structure that accumulates the typical characteristics of each species guaranteed the appearance of primitive cells that were considered the first manifestations of life.

The need to know their structure and functioning, as well as their behavior in different organisms, has led man to study them and today we can recognize their different forms and characteristics in the kingdoms that characterize living matter on the planet.

Methods of study of the cell.

More than two centuries ago, man, in his attempt to learn more and more about nature, undertook the study of cells. Since then, information has been provided about their structures and functions, complementing each other.

The studies on cells that have been carried out so far combine descriptive and experimental methods, using different techniques such as optical and electron microscopy, cell fractionation, cytochemical and auto radiographic techniques and tissue culture, among others.

The human eye, even when using magnifying instruments that allow visualizing components of nature, reaches a limit in which it is not possible to discriminate finer details of these components. This is determined by the possibility that these instruments and the human eye have to be able to separate some structures that are at a certain distance.

For this reason, the observation of cells and their various structures required the use of microscopes. These instruments opened up a new world of observation.

The different types of cells and their internal structure began to be studied, and smaller cells such as bacteria were also discovered, of which only their effects were known. All this through the optical microscope.

However, these devices that use light for imaging reach a limit at which they cannot discriminate particles or other structures that are too small.

The use of electrons as a source of energy opened up a new field in the study of cells and their structures. It made it possible to observe their ultrastructure and allowed scientists to discover the complex organization of matter in them and in other simpler forms. From this moment on, the electron microscope appeared, which allows the observation of those components of living matter so small that it is not possible to observe with the optical microscope. This invention has revolutionized scientific technological development, which, together with modern computerization systems, has made it possible to broaden the field of scientific research.

Study of the cell.

Living matter is characterized, among other aspects, by its structural and functional organization, by its capacity to exchange with the environment in which it develops and by its ability to reproduce itself in replicas similar to the one that gave rise to it. For the latter to occur, the genetic code of each species must be well delimited in the nuclear material of the cells. Only in this way is it possible for the transmission of characters to occur.

Not all cells have their nuclear material organized in the same way and this is what gives them structural and functional diversity and organization.

Every cell is characterized by a cytoplasmic membrane that surrounds and covers it, a cytoplasm where important processes occur, and nuclear material that may or may not be organized in a nucleus delimited by a nuclear envelope.

Primitive cells were characterized by a low level of organization. They lack a precise boundary between the nuclear region and the cytoplasm, they lack a nuclear membrane. The cytoplasm has a low level of organization because there are no specialized organelles and membranous systems as in other more evolved cells. This type of cell is known as **prokaryotic cells**. Bacteria, which will be studied later, are representatives of this type of cell.

Other cells called **eukaryotes** are characterized by a higher degree of structural complexity than prokaryotes. The aspects that make it distinctive and more evolved are the existence of the nucleus as an organelle delimited from the cytoplasm by an envelope formed by a double membrane and the remarkable degree of organization that the cytoplasm possesses by having membranous systems where different organelles are specialized in various functions which allows considering the cell as the smallest living unit. Examples of eukaryotic cells are algae (except blue-green algae), fungi, protozoa such as amoeba, euglena, paramecia, etc., plant cells and animal cells. The following model shows you the general structure of a eukaryotic cell.

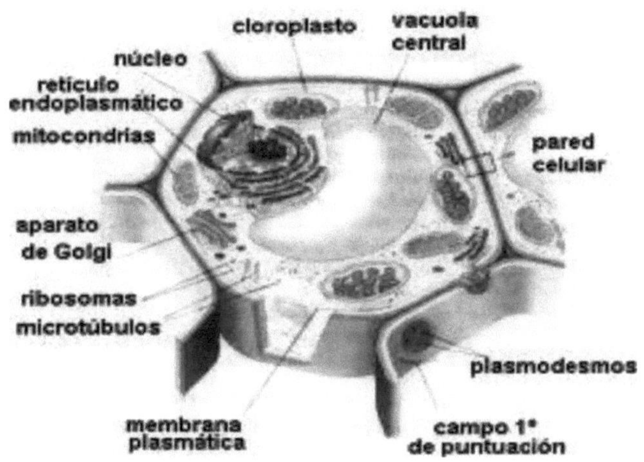

Every cell is delimited by a **cytoplasmic membrane (**also called plasma membrane**)** which makes it possible to maintain the internal content and to carry out exchange relations with its environment acting as a selective barrier. It is a bimolecular layer of lipids between which protein molecules move, which can be found near the inner or outer surface of the membrane. Other times the protein molecules occupy the entire thickness of the membrane. It is known that lipids or fats are impermeable so that substances dissolved in water cannot pass through the membrane. On the contrary, proteins allow the passage of water through them, in these cases when the movement of these molecules occupy the entire thickness of the lipid layer of the membrane they guarantee the permeability of the membrane, they function as hydrophilic regions that guarantee the passage of water and ions. This allows us to understand that it is not possible to think of the existence of rigid or static pores in the cytoplasmic membrane, but rather functional pores or exchange zones caused by the position occupied by the protein molecules in their movement in the lipid layer of the cytoplasmic membrane, which gives it its dynamic character.

Some cells, such as plant cells, in addition to the cytoplasmic membrane, are lined externally by a **cell wall** with a high cellulose content. This also allows the exchange with the medium through specialized zones which are the **plasmodesmata** and the **punctation fields**. When plants are studied, this type of cells will be characterized more extensively. However, the cell wall is not exclusive to plants; other organisms such as bacteria also possess it. When studying the Kingdom Monera you will learn more about it.

Delimited by the cytoplasmic membrane, the **cytoplasm** is found inside the cell and is the place where most of the cellular processes occur. It has a very close structural and functional relationship with the nucleus, which is responsible for the regulation of these processes.

Within the cytoplasm are the organelles that are specialized in different functions and surrounding them is a material called cytoplasmic matrix characterized by a certain degree of complexity due to its enzymatic content, high water content that favors the chemical reactions that occur in it and in which important cellular functions are performed that together contribute to the development of cellular metabolism. In the cytoplasmic matrix, the internal movements of all the components of the cytoplasm take place.

Within the matrix and forming part of the cytoplasm are different structures that have different morphology, chemical composition and functions. These structures, called cytoplasmic organelles or organelles, are in constant movement and are related to each other, especially from a functional point of view.

Below you will learn some characteristics of the main cytoplasmic organelles of the cell so that you can understand why it is the smallest living unit.

Within the cytoplasm there is a membranous complex in which there is a structural and functional differentiation that determines cell dynamics. The **endoplasmic reticulum** is characterized by a system of membranes in which channels and clusters connected to each other are dispersed throughout the cytoplasm. They are involved in the process of protein and hormone synthesis among other components. Endoplasmic reticulum can be smooth or rough; the latter are so called because they have some ribosomes attached to their membranes.

If you observe the cellular structure of the illustration, you will realize that the endoplasmic reticulum corresponds to the nuclear envelope because it is precisely the one that concentrates in a certain area of the cell all the nuclear material and because the reticulum is a membranous complex, a part of it is in direct contact with the nucleus and the outermost part with the cytoplasm.

Other important organelles in cells are **ribosomes,** which are made up of proteins and ribonucleic acid synthesized in the nucleus from where they are transported to the cytoplasm. They can be found associated with the membranes of the endoplasmic reticulum or free in the cytoplasmic matrix. In both cases their fundamental function in the cell is the synthesis of proteins forming hormones and others act in cell metabolism. Some studies have shown that they can also participate in cellular exchange, that is, they are proteins that renew the different molecules that constitute the cellular structures.

The **Golgi complex**, named after the scientist who discovered it (Camillo Golgi, German histologist, 1898), is characterized by flattened sacs resembling arches arranged in parallel to form two or three groups.

These sacs generally appear around another organelle called the **centriole.** The other structures of the complex are formed by vesicles and vacuoles that originate from the flattened sacs.

The main function of the Golgi complex in cells is the synthesis of polysaccharides.

Within cells, processes related to cellular digestion occur, this being understood as the mechanism by which cells degrade the materials they ingest for subsequent use as raw material in their metabolism or in the replacement of their different components. These functions are carried out by **lysosomes,** structures with a high enzymatic content. They are made up of more than forty enzymes that act on proteins, nucleic acids, polysaccharides, lipids and also on sulfates and phosphates.

It is known that all living matter exchanges with the environment and carries out important metabolic processes to release the energy necessary to live. Within the cytoplasm there are organelles called **mitochondria** that have a direct impact on these important processes. Among their components are a large number of enzymes that participate directly in aerobic cellular respiration, a process by which a certain amount of energy is released and used in cellular activity.

It is important to note that the oxygen that enters the cell is fixed in the mitochondria and it is precisely within them where the different stages of aerobic cellular respiration occur, so important in the release of energy and other components such as CO2 that is released into the atmosphere and used by plants in the process of photosynthesis.

Precisely for the realization of photosynthesis, characteristic of photosynthesizing organisms with predominance in plants, appear in plant cells organelles called **plastids** that have a varied morphology according to the function and substances they synthesize. These are the leucoplasts, which are colorless and specialize in storing oils and starches. Chromoplasts, on the other hand, offer plants a variety of colorations. They can store red, yellow, blue, violet, etc. pigments. However, those specialized in storing green pigments with high chlorophyll content are called chloroplasts, which are directly involved in the process of photosynthesis. They will be studied later.

There are other organelles in the cytoplasm such as **microtubules and microfilaments** that play important roles in cells. Microtubules play an important role in cell division, as they are involved in the movement of chromosomes towards the poles of cells in the processes of cell reproduction. They are also involved in color changes in some fish, as they are involved in the movement of pigments through the cytoplasm of specialized cells in their integument.

Their function is also related to the maintenance of the shape and movement of the cell, although they are not rigid structures, but on the contrary, they can modify their position and their components.

Protein microfilaments contribute to the maintenance of cell shape and to different movements that occur in cells, such as the cytoplasmic currents that originate around vacuoles in plant cells and in the formation of pseudopodia as in amoebae.

Other movements occur in the cells for which some organelles called centrioles, cilia and flagella are responsible.

The **centrioles** are part of an area called centrosome located as mentioned above in most cases in the central region of the Golgi complex. The centrioles are formed by microtubules.

When observing some unicellular organisms or other cells in tissues that form different types of organs, the presence of organelles located on the surface of their cells called **cilia and flagella** can be appreciated.

Cilia are characterized by being short and abundant in quantity. Some are arranged over the entire surface of the cell, as occurs in paramecia, whose function is associated with the translational movements of these free-living unicellular organisms. In other cases, in certain areas of the cells that form epithelial tissues, the cilia participate in generating a flow that determines a current in a particular direction, favoring the movement of certain particles that reach these areas. For example, the trachea is lined by an epithelial tissue with cilia that favors the expulsion of any particle lodged in its interior that should not reach the bronchi.

In the following illustration you can see the existence of cilia on the external part of a paramecium. Observe their size and quantity.

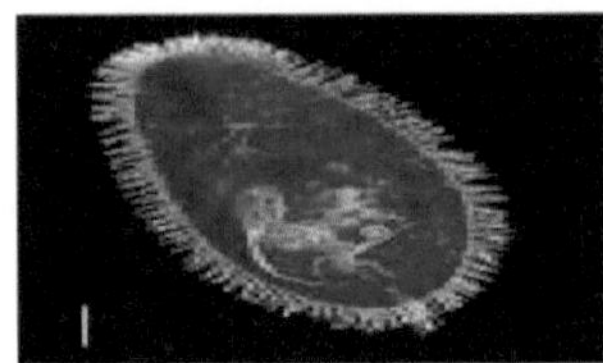

Unlike cilia, flagella are long and flexible in reduced number. For example in some protozoa one appears as in euglena. See the following image:

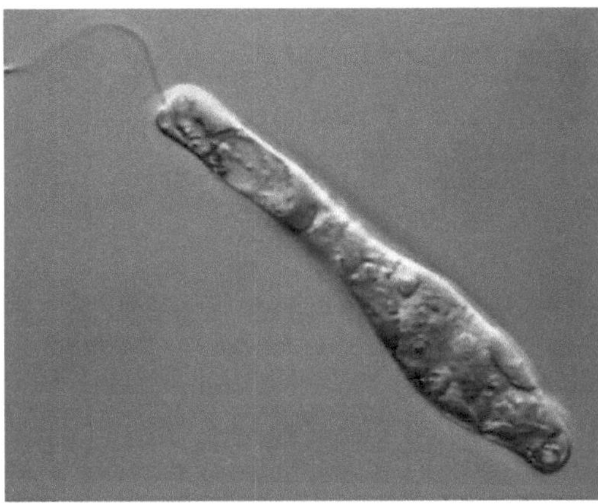

Other cellular components with well-defined functions are the **cytoplasmic inclusions** that constitute accumulations of materials involved in cell metabolism or that are products of it. These include lipids, starches and glycogen, which are necessary energy constituents for the cell.

As you have been able to appreciate the cytoplasm of the cell is very complex because of the number of structures specialized in different functions that are found in it. It is precisely this that offers integrity and makes the cell a dynamic unit.

Another of the cellular structures of relevant importance and distinguishable in the cell by its size is the **nucleus**. It is made up of chromatin, formed by small fibrils and coiled granulations of different shapes depending on the metabolic activity of the nucleus.

The nucleus is delimited by an envelope consisting of two membranes. An outer one, in contact with the cytoplasm, which continues with the endoplasmic reticulum and has ribosomes associated with its surface, and an inner one, in contact with the nuclear contents, which lacks ribosomes associated with it. The nuclear envelope is characterized by interruptions where the inner and outer membranes join, leaving an orifice called nuclear pore that ensures communication with the cytoplasm regulated by some type of activity.

Chromatin consists of deoxyribonucleic acid (DNA) and proteins; the latter are mostly synthesized in the cytoplasm from where they migrate into the nucleus. These proteins are known to be associated with nucleic acids and play an important role in the mechanisms that regulate hereditary characteristics.

When observing the nucleus under the electron microscope, a generally spherical structure can be distinguished, not delimited by membranes and which can usually be found in numbers of 1 to 4, called the **nucleolus**.

It has been possible to appreciate the presence of DNA in the nucleoli, which is involved in the organization of these structures forming part of certain regions of the chromosomes. The material synthesized in the nucleolus migrates to the cytoplasm where it plays a decisive role in protein synthesis.

When the cell is in its reproductive activity, the nuclear envelope disappears, the nucleolus is no longer visible and the chromatin is organized in the form of structures separated from each other to form the chromosomes. The DNA that forms part of these chromosomes has a decisive influence on the transmission of hereditary characters.

Thanks to the possibility of scientific and technological progress, the study and classification of chromosomes in pairs has made it possible to determine the number that characterizes each species. In addition, the study of DNA has made possible the diagnosis of some diseases as well as the clarification of criminal acts.

Unity and diversity of living organisms.

After the first cells originated, they evolved over a few billion years to produce the rich biodiversity that characterizes our planet today.

When you look around any place, the first thing that strikes you is the diversity of organisms that exist. It is not known exactly how many species of organisms may exist, but most biologists estimate that there are at least 5 to 10 million. Each of these organisms has its own unique life history, its own way of obtaining energy, its own place in the world of living things.

Living organisms present common characteristics that offer unity, they are all born, develop and die. They all perform common functions such as nutrition, obtaining energy, reproduction, etc. However, there are aspects that make all living beings diverse: their shape, size, color, adaptation to habitats, forms of nutrition, reproduction, respiration, locomotion, etc. It is therefore a reality, in nature there is unity and diversity of the living world.

Due to their characteristics, organisms can be organized in a single cell and are therefore called **unicellular**, while others are considered **multicellular** because they are formed by more than one cell.

When considering this form of structural organization of living organisms, it must be understood that there are different levels of organization of living matter.

Thus, it is considered that **tissues** are formed from a **cell** when they are grouped together with similar structures adapted to the same function. The groups of tissues that are associated to structurally guarantee specific functions form the organs. These in turn related to each other for the performance of the same function contribute to the organ systems. The set of organ systems closely related to ensure the integrity of functions constitute the **organisms**. All of them are grouped into **populations** that are organisms of the same species. When under natural conditions several populations develop in the same ecosystem, **communities** are formed. The set of all the communities on the planet constitute the **Biosphere**, which is the sphere of life.

Consequently, living matter is organized as follows:

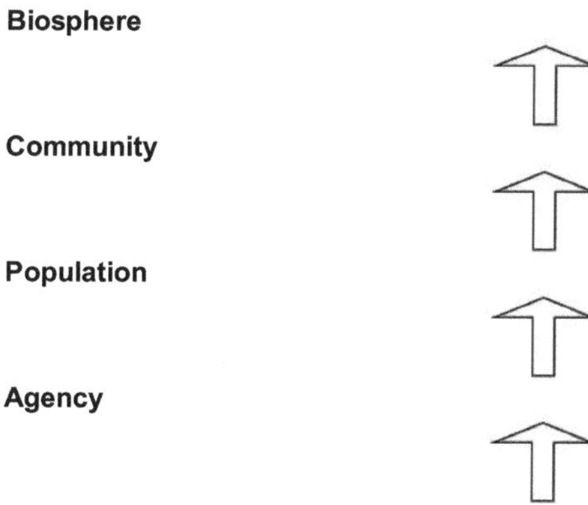

Cell

When you look at the illustration below you will appreciate exactly how living matter is organized on the planet.

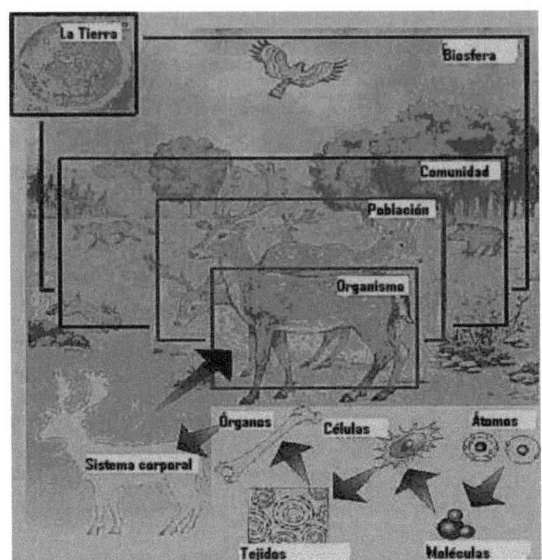

This enormous diversity of species inhabiting Planet Earth forced biologists to face the task of ordering them, in order to facilitate their study. This implied classifying and naming each one of them.

Thus were born two specialized branches of the biological sciences, **zoological taxonomy and zoological nomenclature.** The former is responsible for establishing the theories and practices relating to the classification of animals, and the latter for giving distinctive names to each of the groups recognized by classification.

Nomenclature.

In different countries, and in regions within a country, there are multiple ways of naming the same organism, which constitute the ***common names***.

The domestic dog is given the following names in eight languages:

Francés - chién

Alemán - hund

Italiano - cane

Inglés - dog

Polaco - pies

Ruso - sabaka

Danés - hond

Hebreo - kelev

This problem has been solved with the use of **scientific names,** unique for the whole world and written in Latin or Latinized. Each individual species is given two names (**binominal nomenclature**). The first word is capitalized and corresponds to the **genus**, and the second (which is written in lower case) is the **specific epithet or adjective**, usually descriptive or geographical. This is usually followed by the surname of the scientist who described and named the organism, or simply an abbreviation or the initial letter, if it is well known. For example: the wolf ***(Canis lupus, Linnaeus).***

This form of naming was established in 1758 by the Swedish naturalist Linnaeus, the founder of modern taxonomy. He used Latin names because the scholars of his time communicated in this language.

The following examples demonstrate the above: wolf (**Canis *lupus*)**, coyote ***(Canis latrans)***, domestic dog ***(Canis lupus familiaris*** L) and common fox (**Vulpes vulpes**).

Ranking

Zoologists classify individual animals at the basic species level, which is the only such category that can be considered in nature. The higher categories are assemblages of groups of species.

A ***species*** is composed of groups of natural populations that actually or potentially interbreed with each other, share a common gene pool, and are reproductively isolated from other similar groups. Species that are clearly related by sharing important characteristics are grouped into a ***genus.***

To construct the hierarchical classification, one or more genera are grouped into *families,* families into *orders,* orders into *classes,* classes into *phyla or divisions, and* these into *kingdoms.*

The groups of organisms included in these seven main categories, at any level of hierarchy, are called ***taxa.***

To allow further subdivision, the prefixes *sub- and supersedes* any category can be added. In addition, in complex classifications, special intermediate categories such as ***branch*** (between kingdom and phylum), and ***tribe*** (between family and genus) can be used.

The following is the classification of gorilla and man as an example.

It can be seen that they are included in the same categories up to the suborder level.

	Gorilla	**Man**
Reino	Animalia	Animalia
Phylum	Chordata	Chordata
Subphylum	Vertebrata	Vertebrata
Class	Mammalia	Mammalia
Subclass	Eutheria	Eutheria
Order	Primate	Primate
Suborder	Anthropoid	Anthropoid
Family	Pongidae	Hominidae
Genre	Gorilla	Homo
Species	*Gorilla gorilla*	*Homo sapiens*

For these reasons it can be understood that the specialists in classifying (taxonomists) are based on similarities and differences that organisms present, therefore many classifications have been recorded based on the elements that science has contributed up to a certain moment and on the criteria assumed by certain authors.

At the beginning there were the Plant and Animal kingdoms, later they discovered differential characteristics between them that determined other classifications.

At present, the classification of R. H. Whitaker, who established five kingdoms, is assumed. This is the most recent, however, as discoveries and scientific studies progress, the limitations that this classification may present will be perfected. The criteria followed by the author groups the organisms as follows: **Kingdom Monera,** where all unicellular organisms with prokaryotic cells have been included; **Kingdom Protista,** which comprises unicellular organisms isolated or forming colonies, whatever type of nutrition or movement they present. Their reproduction is mainly asexual with eukaryotic cells; **Kingdom Fungi,** which groups organisms that have generated controversy among taxonomists due to their great heterogeneity, which is why their morphological characters, the peculiarities of their reproduction and biochemical studies of their metabolism have been taken into account to consider that they form an evolutionary line of their own within the multicellular eukaryotic organisms; **Kingdom Plants,** which includes all eukaryotic multicellular organisms that possess cell wall and pigments in cytoplasmic organelles with autotrophic nutrition thanks to the process of photosynthesis, live fixed to a substrate and their reproduction is mainly sexual and **Kingdom Animalia** (Animals), where all eukaryotic multicellular organisms that do not possess chlorophyll and have adaptations for movement on their own, are heterotrophic and their reproduction in most cases is sexual, were grouped.

In the following chapters you will study in detail the characteristics of the Animal kingdom.

Self-monitoring questions

1. Summarize in a table, the main events that took place so that life could arise on planet Earth.
2. How important is the discovery and use of microscopy to the study of life on Earth?
3. "The cell is considered the structural and functional unit of all living organisms." Argue the above statement.
4. State the fundamental differences between prokaryotic and eukaryotic cells. Cite examples of each.
5. Explain the dynamic character of the cytoplasmic membrane of cells.
6. Make a table where you can summarize the functions performed by the different cytoplasmic organelles of the eukaryotic cell.
7. Summarize the importance of the nucleus in the eukaryotic cell.
8. Express the criteria that make possible the unity and diversity of the living world.
9. Value the importance of classifying organisms according to the characteristics that are common to them and that make them different from others.
10. Name the five kingdoms into which living organisms are grouped for study.

CHAPTER 2: ANIMAL KINGDOM

Next you will study a living kingdom whose characteristics are heterotrophic and therefore differ considerably from plants, despite having common elements that you already know.

To begin with, **what is an animal**?

Animals are made up of numerous cells, whose cellular pattern is eukaryotic, so they are considered multicellular. In addition, these cells lack cell walls and plastids, unlike plants. The lack of chlorophyll in animals determines the type of nutrition, which is why they are considered to have heterotrophic ingestive nutrition, with some exceptions, such as some parasitic worms like the tapeworm, whose nutrition is heterotrophic absorptive and which has arisen as an adaptation to parasitic life. In animals reproduction is typically sexual. This does not mean that there are no individuals with asexual reproduction and that in some cases the alternation of generations is manifested.

Other important characteristics that animals present is that in all of them there is an embryonic development and locomotion that allow them to find ecological resources to live in harmony with the environment. With these elements you will be able to understand why, being animals so diverse, they present characteristics in which their unity is evident, so animals are:

Multicellular organisms, consisting of eukaryotic cells, lacking cell wall and photosynthetic pigments, with heterotrophic ingestive nutrition, with typically sexual reproduction, embryonic development and locomotion.

In order to understand the characterization of the different groups of organisms studied in the kingdom Animalia, it is first necessary to become familiar with some concepts used in zoology.

Symmetry is that which refers to the distribution of the different parts of the body with reference to an axis.

Thus, an organism is considered to have **radial symmetry** when the parts of the body are arranged around a central point like the spokes of a wheel. That is, they can be divided by different planes resulting in equal halves. Most animals with radial symmetry are sedentary or very little mobile, which allows them to receive stimuli from all directions of the environment. Examples of animals with radial symmetry are Cnidarians and Echinoderms in the adult state.

Bilateral symmetry is that in which the animal can be divided into two very similar halves only by one plane. Animals with bilateral symmetry are more complex, have faster locomotion movements and show a higher degree of cephalization (greater development of the nervous system and sense organs). Examples of animals with bilateral symmetry are Nematodes, Platyhelminths, Annelids, Arthropods, Mollusks and all Chordates (man is an animal with bilateral symmetry).

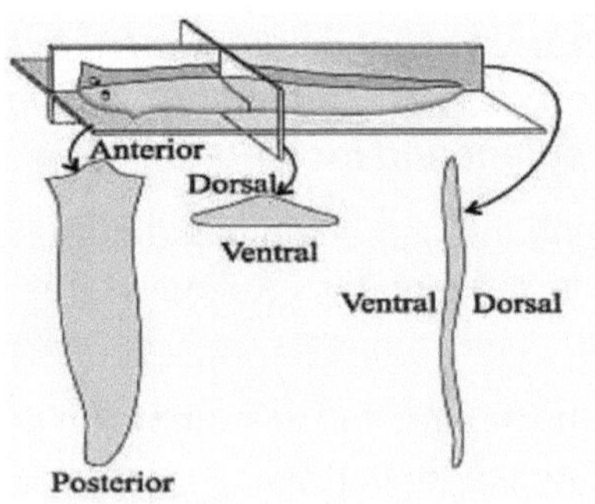

When there is no plane by which the animal can be divided into equal parts, they are said to be **asymmetrical**, i.e. there is no well-defined symmetry.

Examples of animals that are asymmetrical are some sponges.

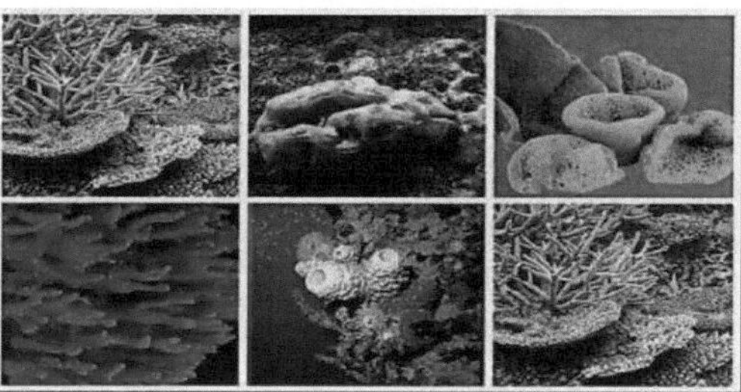

Another important element when classifying animals is the **type of body cavity.**

Most animals have their body structure from the development of three embryonic layers. The outer layer, called the **ectoderm,** gives rise to the outer covering of the body and the nervous system. The inner layer, or **endoderm**, covers the digestive tract. The middle layer, the **mesoderm**, extends between the ectoderm and endoderm and gives rise to most of the body's structures, including muscles, bones, circulatory system, etc.

Within the animal kingdom, we find groups that present the germinative layers one after the other, i.e., ectoderm, mesoderm and endoderm without a cavity between them (e.g., flatworms, liver flukes, planaria, dog tapeworms, etc.).

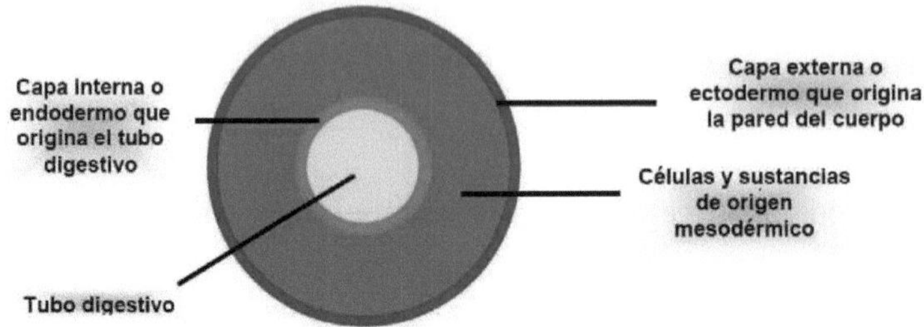

In others the cavity is located between the mesoderm and the endoderm and are called **pseudocellomates** such as nematodes.

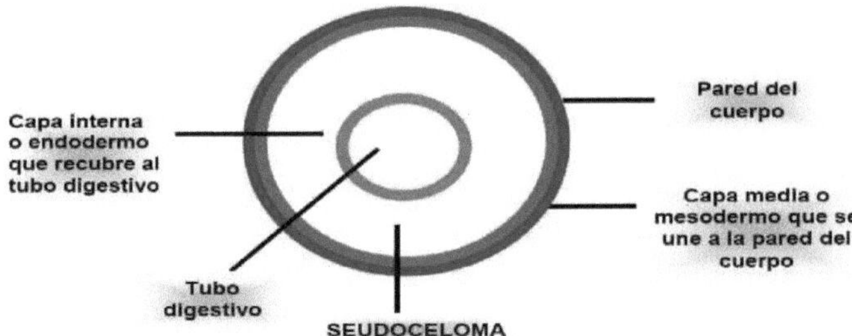

In most taxonomic groups the cavity is located within the mesoderm itself and are called **coelomates** such as annelids, mollusks, arthropods, echinoderms and chordates.

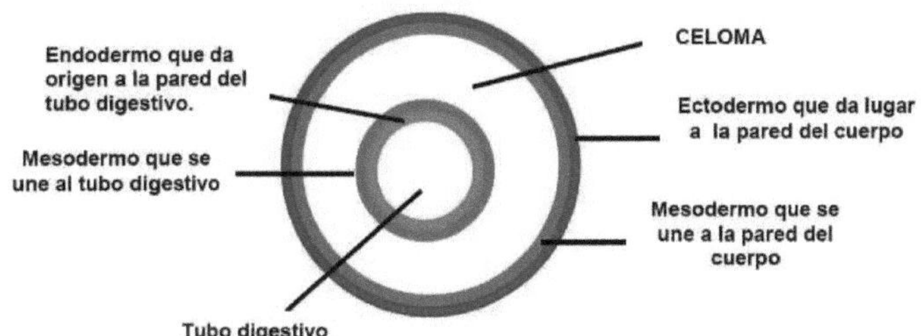

Another important characteristic to take into account when describing the anatomy of animals is that of their **tissues.**

Animal tissues have characteristics in common with plant tissues, since they are groups of cells similar in structure and function; however, in the case of this kingdom, their structure and function have other particularities.

In animals, their bodies are made up of four **different** groups **of tissues**: **epithelial, connective, muscular and nervous.**

Epithelial tissues constitute the superficial layer of the skin, they cover the mucous and serous membranes of some organs and form glands. Epithelial tissues can be formed by a single layer of cells called simple epithelium or by more than one layer called in this case stratified for forming layers of cells.

Let's see some particularities of this tissue that will allow you to understand the characteristics of some organs that you will study later. The cutaneous epithelial tissues constitute the superficial layer of the skin, the membrane that lines the oral cavity and part of the throat. This tissue is of the stratified type. This cutaneous epithelial tissue plays an important role in the defense of the organism as it protects it from the action of different external agents such as chemical, thermal and mechanical changes. Heat is also released through it. In the following chapter you will be able to better understand these aspects by studying the human organism and with it the skin, the respiratory, digestive and other systems.

It has also been said that there is glandular epithelial tissue, which is the fundamental tissue of the glands and specializes in producing or secreting special substances with different functions in the organism.

Another form of epithelial tissue is the one that forms the serous membranes, that is, those that cover the cavities of the organism. For example peritoneum, pleura, pericardium, etc. (These aspects will be deepened in the following chapter).

Connective tissues as their name indicates their function is to unite since in the intercellular spaces of their tissues there is a well-defined intercellular substance according to the function that each variety of connective tissue plays in the organism. This group includes the blood and lymph that circulate throughout the body connecting all organ systems; the lax fibrous connective tissue that joins the blood vessels constitutes layers between organs and is also found below the subcutaneous layer.

The lax fibrous connective tissue fulfills in the organism the function of support, defense and trophic. The supporting function is carried out by the framework of the organ, giving solidity and elasticity. The defense function is carried out by macrophages, cells that actively participate in the fight against microbes that cause diseases introduced into the organism. The trophic function is associated with the participation in the feeding process of the tissues of the various organs, in other words, through the intercellular substance of this tissue the food substances that reach the various organs of the animal's body are transmitted.

In the lax connective tissue there is another very important variety, the adipose tissue located under the skin, around the vessels and many organs. The adipose tissue has a trophic function, that is, it stores a large amount of substances with energy reserves in the form of fat that the organism can use when it is necessary. For example, some bears hide until spring in the harsh winter and their metabolism is reduced to a minimum, in which case they are sustained by the energy reserves accumulated in the adipose tissue. In the same way, any organism uses these reserves if it needs them. When this does not happen, the existence of a large amount of adipose tissue causes obesity.

Adipose tissue also helps to protect vessels and other organs from the effect of some shocks or mechanical effects in general.

The supporting connective tissues are made up of dense fibrous, cartilaginous and osseous connective tissue.

Dense fibrous connective tissue constitutes the tendons, ligaments and the cutaneous base (skin proper). This tissue generally has a supportive function.

The cartilaginous tissue has a diversity of components in its intercellular substance, so its function depends on it and we can find it in the larynx at the junction of the ribs with the sternum and the cartilages of most of the joints. Another form of tissue forms part of the pinna and the epiglottis characterized by many elastic fibers in its basic substance. Finally we find this type of tissue in the intervertebral cartilages and in the insertion of tendons with bones.

Bone tissue is another form of connective tissue made up of bone cells called osteocytes and intercellular substance which possesses mineral salts that give solidity to the tissue. This tissue forms the bones that make up the internal skeleton of vertebrate organisms in general, except for cartilaginous fish. **Muscular tissues** are characterized by the property of contracting and relaxing. They include smooth and striated muscle tissue. In the case of the former, the fundamental characteristics are the existence of elongated fibers with a cytoplasm containing a rod-shaped nucleus. In its cytoplasm, called sarcoplasm, special structures are distributed, the contractile filaments or myofibrils.

This type of tissue is characteristic of the walls of organs such as the intestine, bladder, uterus, stomach and others. It also forms part of the structure of the walls of blood vessels and in the skin.

The striated muscle tissue constitutes the skeletal muscles, the cardiac muscle and some internal organs such as the soft palate, the tongue, etc.

Its fundamental structure is the muscle fiber characterized by a large number of nuclei and an enveloping membrane in the sarcoplasm. The myofibrils are also located, which are not uniform, presenting dark and light bands that are associated with muscle contraction. The existence of several nuclei guarantees the efficiency of the muscle fiber in the contraction processes. Muscle fibers are grouped in fascicles separated from each other by layers of lax fibrous connective tissue.

In the case of the striated muscular tissue of the heart, some particularities appear that will be studied in the following chapter.

The nervous tissue is the basic element of the nervous system, which regulates the processes that take place in the organism and ensures the mutual relationship with the external environment due to the properties inherent to this tissue such as excitability and conductivity. In other words, the different stimuli that act on the organism provoke excitability, these are conducted in the form of impulses by the nervous tissue.

The characteristic nerve cell of this tissue is the neuron, which consists of a body from which axons and their endings emerge. The transmission of the nerve impulse from one cell to another occurs through a mechanism called neuronal synapse. The transmission of the impulse to a muscle through a plate specialized for this function is called neuromuscular synapse.

The Kingdom Animalia includes different branches based on their particular characteristics. In this chapter we will describe the general characteristics of some of the groups belonging to these branches and their representative specimens for their importance in nature, health and economy, which will prepare you for your performance as a primary teacher.

Phylum Porifera

The term Porifera comes from two Latin words **porus**, meaning pore and **ferre**, meaning to carry, these animals are commonly called sponges and are mainly marine, although some are freshwater, sessile in adult state and generally inhabitants of hard substrates; presence of many pores; they are asymmetric; they lack tissues, organs and organ systems; they present different types of cells, but their cellular differentiation has not followed the pattern that is common for other animals. Their body is organized around a system of water channels, in correlation with the sessile life of organisms; the inner surface of the body covered by collar cells or coanocytes; the outer part covered by cells called pinacocytes and traversed by cells called porocytes, skeleton of organic spongin fibers, siliceous or calcareous spicules or both; feeding, gas exchange and elimination of wastes depends on the flow of water through the body; intracellular digestion; asexual reproduction by buds; also sexual; free-swimming ciliated larvae.

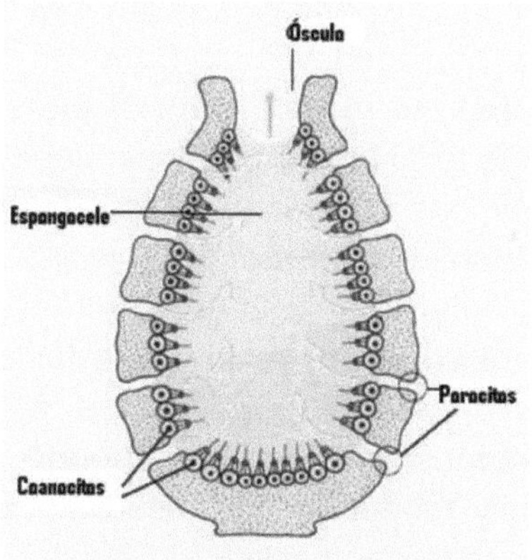

Natural sponge specimen

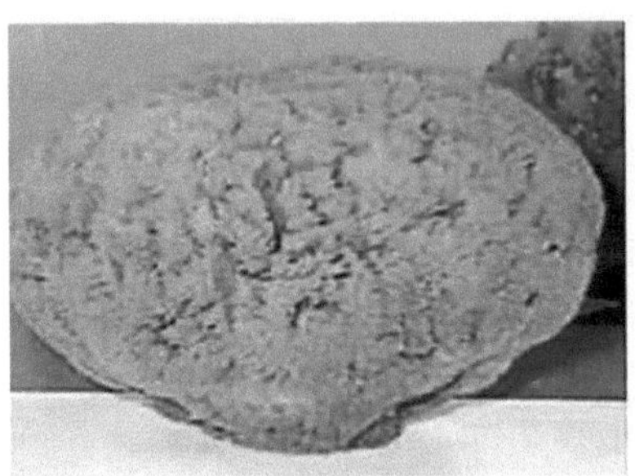

Cross section of a sponge

The body of the sponges has radial symmetry and is composed, in its external part *(**Ectoderm**)*, by flat connective tissue cells very contractile, although with amoeboid movements (no type of musculature is seen in any porifer). The inner part is formed by cells called ***coanocytes*** equipped with a flagellum, they have a dual function, the food and the maintenance of the water flow from the outside of the sponge to its interior. Between the two layers is a more or less dense, gelatinous substance *(**Mesoglea**)* containing various types of cells, the mobile *(**amoebocytes**)* which are responsible for communicating the outer layer with the inner layer and transporting food, others (***sclerocytes***) are responsible for secreting a kind of skeleton to support the soft body of the sponges, formed by calcareous or siliceous *spicules* or spongin fibers (a protein substance chemically related to the hairs and horns of mammals); they can also give rise to germ cells.

The adult sponge is a sessile animal, unable to move, relying for its nutrition on systems of channels, chambers and sparsely organized flagellated cells. Given the absence of nerve cells (indicating the low organization of sponges), there must be some kind of communication between the different cell types, possibly by chemical diffusion, for this aggregate of different cell types to function as a single animal. Its ectoderm is perforated by numerous pores through which new water, supplied with oxygen and food particles, constantly enters its interior cavity, a **spongocele** also called atrium, once used, it is expelled to the outside through the ***osculum***. The water current is produced by the synchronized movement of the flagella of the coanocytes that line the atrium.

The classification of sponges is based on the skeletal materials that provide support to the sponge, i.e. the spicules. There are three classes of porifera (sponges):

1. **Calcareous sponges**. They include littoral species, whose spicules are exclusively calcareous.
2. **Siliceous sponges**. Includes deep-water sponges, whose siliceous spicules, separated or joined together, form skeletal networks of great elegance.
3. **Corneous sponges**. Includes sponges whose skeleton has variable shape, and their spicules are siliceous, or are composed of spongin (corneous elastic protein substance). Sometimes the spicules are also composed of both components at the same time (silica and spongin).

The reproduction of sponges is very variable. Almost all reproduce sexually, although all individuals can produce eggs and sperm indistinctly, as they occur separately there is never a self-fertilization, but we can refer to a cross-fertilization. They also reproduce asexually, by means of reproductive bodies called *gemmules,* which either split off, becoming an independent animal, or remain attached to the mother sponge, thus forming sponge colonies.

An important characteristic of sponges is their ability to regenerate. They can suffer damage and recover the parts of their body that were affected or lost.

The morphological characteristics of sponges have allowed man to use them for economic purposes. Their siliceous or calcareous skeleton and the spongin present in them offer multiple uses, which is why there are industries specialized in their treatment and subsequently commercialized with dissimilar uses, for example in cosmetic accessories, as packaging supports, in furniture upholstery, etc.

The care of the seabed and sponge nurseries is essential for the conservation of these species and for their subsequent use.

Phylum Coelenterata or Cnidaria

The animals of the phylum Cnidaria (gr. ***Knidé,*** nettle) are characterized because in them the cavity of their body represents the digestive tube; it is at the same time, digestive cavity and general cavity, to which alludes the name **celenterates**. Coelenterata (*gr.* ***koilos,*** hollow*, and* ***enteron,*** intestine), as they are also known. This digestive cavity, in them called gastrovascular cavity, communicates with the outside through a single opening that serves as mouth and anus, and is surrounded by tentacles, provided with stinging cells or nematoblasts, which they use as a defense mechanism.

The body of coelenterates presents a level of organization in tissues, although these, in general, do not constitute organs, except for some sensory structures present in some species. In these animals two layers of cells are observed: the ***epidermis*** *and* the ***gastrodermis,*** *and* between them there is a layer devoid of tissues, although there may be cells that have migrated from the epidermis or gastrodermis; this layer is called mesoglea (gr. *mesos,* medium, and *gloia,* gelatin), which does not correspond to the mesoderm, so the coelenterates are diploblastic.

The mesoglea can have different thickness and consistency; sometimes it contains amoeboid cells, in which case it is called cenenchyma.

In bad waters (jellyfish) they constitute the thick mass that makes up the most voluminous part of their body.

Symmetry in these animals is radial (tetramer or polymer) or biradial.

Radial symmetry is a primary symmetry in them, but in the more complex groups this symmetry becomes biradial, because the mouth and the aboral end are located at two opposite points of a main axis, around which the different parts of the body are arranged concentrically.

All coelenterates are aquatic and mostly marine; only ***Hydra*** *and* other representatives are freshwater.

About 10 000 species are known, generally inhabiting coastal regions, especially in tropical seas, although there are species in all seas, including the glacial ocean. Most of the species (6 000) are benthic and many are pelagic. Benthic species are abundant on the rocky coasts of tropical waters.

Types of coelenterates according to their structural organization.

There are two main types of coelenterates: those with a cylindrical shape, called ***polyps,*** which are sessile, and those with an umbrella shape, the ***jellyfish,*** which can swim freely, both as solitary or colonial individuals.

The polyp is fixed and sac-like with the opening facing upwards. Certain polyps are capable of making a calcareous internal skeleton that persists after death and contributes to the formation of coral reefs. Many species form colonies of polyps, often polymorphic.

Coral polyps

The jellyfish is free-living and has the ability to move in the water. It has the shape of an umbrella, called umbrella (from the Latin umbrella meaning umbrella) from the edge of which hang a series of tentacles. The mouth is on the underside, in the center of another group of tentacles called manubrium. Jellyfish have organs of balance and also specialized organs to capture light. Species include coral, hydra, jellyfish and sea anemone.

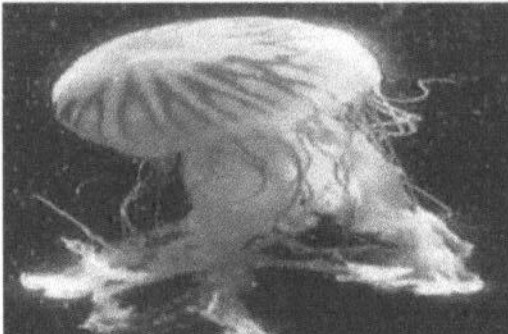
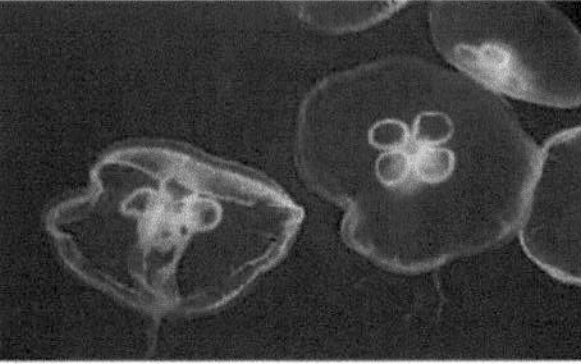

Jellyfish

Both polyps and jellyfish possess, especially in the tentacles, specialized cells called cnidoblasts that contain a vesicle (cnidocyte)

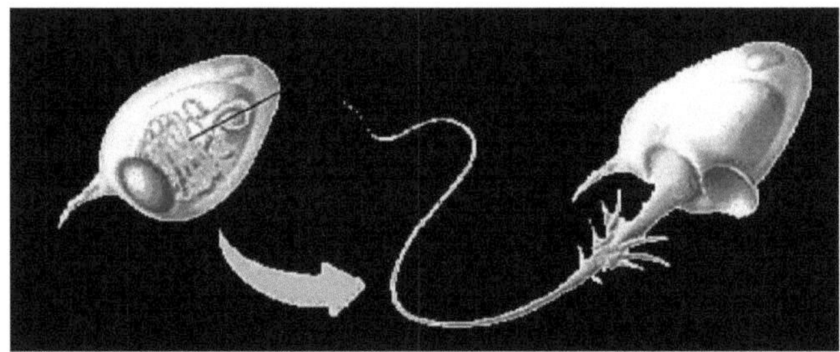

filled with toxic fluid, which they inject into their prey by means of a filament, or cnidocyte, that projects outward when stimulated; Depending on whether it is a penetrating, enveloping or binding cnidoblast, the filament inoculates stinging liquid into the animal that has brushed against the cnidocilium or coils around it. The stinging produced by this substance is felt on the skin when bathers are attacked by jellyfish or bad water on the beach.

Observe in the figure the structure of the cnidoblasts.

The reproduction of coelenterates occurs by alternation of generations. If you look at the figure below you will see that there are two phases.

In this polyp colony, asexual generation consists of two types of polyps: those specialized for feeding and those for reproduction. From the reproductive polyps, male and female jellyfish are released, producing sperm and eggs. When fertilization occurs by the union of both gametes, an egg is formed, which later transforms into a larva called a **planula** that swims and attaches to a substrate, develops and gives rise to a new generation of polyps by asexual reproduction.

Because there are two reproductive phases, sexual and asexual, it is known as alternation of generations, which, as you can see, is not the same as the one already studied in plants.

Look at the illustration below to understand the stages of this alternation.

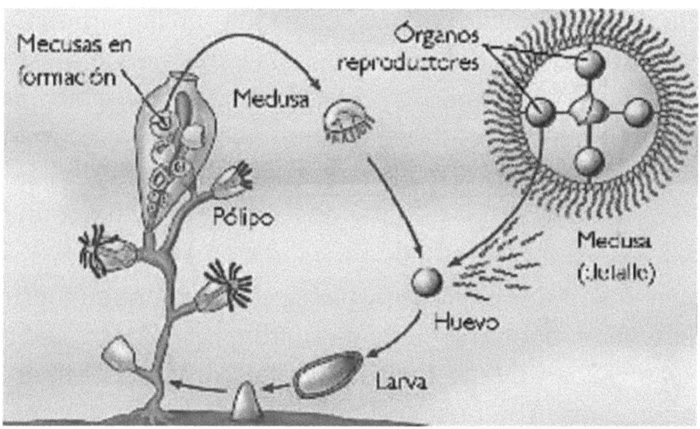

As can be seen, the specimens belonging to the coelenterates are diverse in terms of forms and lifestyles, but there is an important characteristic: some live solitary and some can form colonies. A colony begins with a single individual that is produced asexually by budding, but instead of separating from the parent, the bud remains attached to it and in turn forms more buds. Several types of individuals can be generated in the same colony, some specialized in feeding, others in reproduction and others in defense.

The specimens studied in this Phylum are of great importance. Among them we cannot fail to mention the corals because although some can capture prey, many tropical species depend for their nutrition mainly on photosynthetic dinoflagellates that live inside their cells.

In warm marine waters, almost every square meter of the bottom is covered with corals, brightly colored and forming reefs.

In the South Pacific the reefs found are the remains of billions of microscopic calcareous skeletons secreted in the past by coral colonies. Only on the surface of such reefs are living colonies found, which contribute their own skeleton to the calcareous rock in formation.

In Cuba it is characteristic to find coral reefs that provide great importance in the platform where important marine species develop. The colorful colors and diversity of species, determined by the geographic location of the Cuban archipelago, the characteristics of the waters and its insular platform, constitute an important economic line from the tourist point of view. Tourists interested in this topic visit diving zones predetermined by marine biologists, which makes it possible for the number of those curious about this topic to increase, increasing the income of important sums of money when visiting Cuba, which is used in the general development of the country.

Coral reefs constitute barriers that help maintain the characteristics of the ecosystems of our platform, which implies the need for their care and conservation, both to prevent them from being damaged by water pollution, as well as by the indiscriminate extraction of their species for other purposes. The passage of meteorological phenomena such as tropical cyclones has caused disasters in these areas, which has led to the treatment of these areas by specialists and the wait for their stabilization and reproduction of their species.

Phylum Platyhelminthes.

The name **Platyhelminthes** comes from the Greek word **platy**, which means flat, and **helminthes**, worms, so all animals studied within this phylum are flat worms, with bilateral symmetry, they are acelomates, triploblastic since they present the three germinative layers, i.e., ectoderm, mesoderm and endoderm Currently numerous species of flatworms are known.

Of these, those that are free-living and have a soft body, such as the **planarians**, are grouped in the class **Turbellaria**; others, such as the liver fluke, a parasite of man and domestic animals, are studied in the class **Trematoda;** and finally, the class **Cestoda** also includes parasitic flatworms, in this case with an elongated and segmented body, such as the tapeworms.

The Phylum Platyhelminthes comprises an important group of free-living or parasitic animals that inhabit both marine and freshwater environments or as parasites of vertebrates or invertebrates. In Cuba, with the exception of the parasitic forms of terrestrial vertebrates, the other types are relatively little known and still deserve further study.

Cuban parasitic flatworms belong to the ***Cestoda*** *and* ***Trematoda*** classes**,** they inhabit all terrestrial vertebrates, causing great damage especially in domestic animals and much less in wild animals of the national fauna, being bats and birds, the major hosts of these parasites.

In general, flatworm species are small in size and vary from microscopic dimensions up to about 6 to 7 cm, although some tapeworms can reach several meters in length. Numerous flatworms are parasitic on domestic animals and humans.

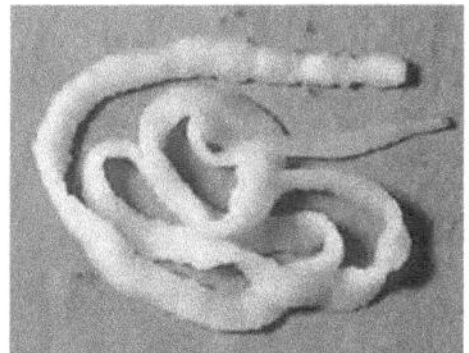

I had

Many flatworm species are cosmopolitan, that is, they are widely distributed throughout the planet; others are specific to each country; and there are many that are not yet known and it is hoped that the increase in scientific development in all countries of the world will allow their study and knowledge.

The most important characteristics of this Phylum are:

- They have bilateral symmetry with well-defined anterior and posterior ends, which is advantageous, since having a well-defined anterior end, their locomotion is directed in that direction. They also have sensory organs in the part of the body that is exposed to the environment. The existence of a rudimentary head is therefore evident, which represents the beginning of cephalization.
- They have well defined three germinative layers. In addition to an outer epidermis and an inner endodermis, flatworms have an intermediate layer of tissue that develops from the mesoderm.
- They are the simplest organisms in which there are well-developed organs consisting of two or more types of tissues. Among their organs are a muscular pharynx for ingestion of food, eyespots and other sensory organs in the head, a simple brain, and complex reproductive organs.
- A simple nervous system consisting of a brain with two masses of nerve tissue called ganglia connected to two nerve cords that extend the entire length of the body.
- Excretory structures called protonephridia, which terminate in collecting cells called flammiferous cells.
- A gastrovascular cavity in most species with a single opening, the mouth, usually located in the middle of the ventral surface.
- Their size varies up to several meters, for example, tapeworms can be up to 15 meters long.

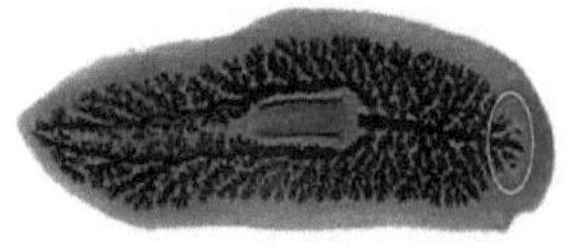  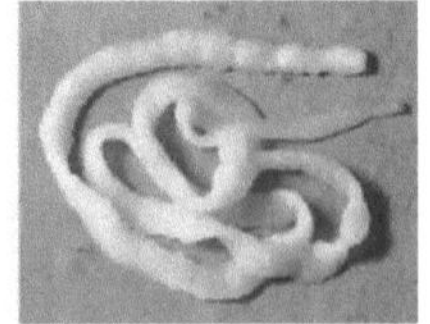

The parasitic flatworms represented by tapeworms and staves are quite adapted to their way of life. They have suckers or hooks to attach themselves to the host. Their bodies are resistant to digestive enzymes secreted by their hosts. Many have complicated life cycles that allow them to change hosts. To further ensure the survival of the species, these worms produce large numbers of eggs.

Among the flatworms there are three fundamental classes: The **Turbellaria class** in which the representative specimen is the **planaria**. Free-living organisms, mainly marine and some terrestrial, living in mud. The body covered by a ciliated epidermis, with carnivorous nutrition form that prey on tiny invertebrates or dead organisms. The **class Trematoda** represented by the **staves** are parasitic with a wide range of vertebrate and invertebrate hosts that may require intermediate hosts such as aquatic snails and possess suckers for attachment. The **Cestoda class** with its body formed by fragments (proglotis) where **tapeworms** are its representative specimens characterized by being parasites of vertebrates, with complex life cycle with one or two intermediate hosts; they have suckers and some hooks on the head (scolex) for attachment to the host, the eggs are produced inside the proglotis which are released, they do not have digestive system.

These parasitic organisms have their life cycle in such a way that their eggs are exposed to the environment and therefore vegetables and other vegetables where there are aquatic snails can be contaminated with staves that produce serious hepatic complications. On the other hand, tapeworms, in part of their life cycle, develop in the musculature of cattle, so that when we eat their contaminated meat we can easily become infected. For this reason, it is necessary to cook food hygienically and with the rigor required.

This also implies applying control measures on crops and livestock to prevent the proliferation of diseases caused by these flatworms.

Phylum Nematoda.

The name Nematoda comes from the Greek word **nema**, which means thread, called by some authors nematelminthes, constitute the most numerous and important phylum of aschaelminthes; among their 10 000 described species, some represent the most abundant metazoans with the widest geographical distribution; Suffice it to say that in an area of one square meter of the muddy bottom of the coasts of Holland more than 4 000 000 specimens were counted, and it has been calculated that in the surface layer of the sand of some beaches there may be hundreds of millions, in addition, many species are parasitic and attack all groups of plants and animals, including man, crop plants and domestic animals, which is why they are of great economic interest.

Nematodes are essentially distinguished by having a cylindrical body, not segmented, covered by a thick cuticle, without suckers, with a constant number of somatic cells, the muscular system consisting only of longitudinal fibers; presence of **pseudocelium**; they lack the parenchyma that exists in flatworms and nemertines; the digestive tract is provided with independent mouth and anus and located at opposite ends of the body; the excretory system has no flammiferous cells; the sexes are generally separate and their spermatozoa devoid of flagella.

Nematodes live in a wide variety of environments, many are free-living and are found in the sea, in fresh water and on land; from the polar regions to the tropics, in deserts as well as in high mountains. Marine species are generally benthic and are found in aquatic sediments, fixed to the substratum, especially on organic-rich bottoms, and are found both in coastal regions and in the great depths of the ocean.

Freshwater species are found at the bottom of lakes and rivers, in pools and even in hot springs, where the water temperature can reach more than 500C.

Terrestrial species essentially inhabit wet sites and can form huge populations among surface soil particles, especially those surrounding plant roots.

Some species inhabit different parts of plants, such as roots, leaf axils, fruits or seeds. Parasitic species show all degrees of parasitism and live inside the tissues or organic cavities of both plants and animals. Parasites of man, for example, are found in the intestine (whipworm, etc.); in the muscles ***(Trichinella** larvae),* in the subcutaneous cellular tissue (Medina filaria), in the lymphatic or blood vessels ***(Wuchereria)***, etc.

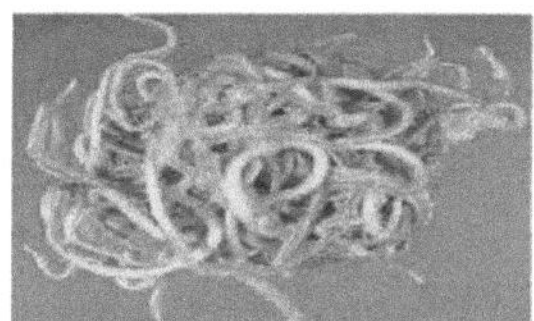

Intestinal worm

Pinworms are the most common worms in children that lodge in the large intestine. The females of these parasites usually migrate to the anus in the evening to deposit their tiny beads. The irritation caused by the presence of these nematodes in the anus causes people to scratch themselves and this disperses the tiny beads. In the nails or in the environment they can be spread and brought to the mouth when eating food or any other manipulation. In this way we become infected. During a heavy infestation, discomfort, irritation and lesions appear in the intestinal wall.

For this reason it is important to take a series of measures to avoid being parasitized by these animals;

- Wash hands before eating food.
- Wash food well before eating.
- Do not walk barefoot.
- Do not wear other people's underwear.
- Boil plain clothes.
- Defecating in latrines.
- Boil or chlorinate water for drinking.
- Cooking food well, etc.

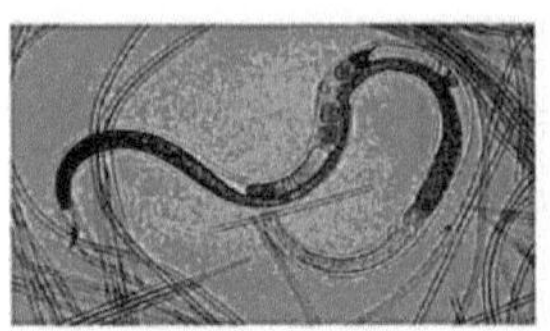

Las fotos muestran un nematelminto de vida libre entre las yerbas acuáticas y un grupo de ellos parásitos del intestino del hombre.

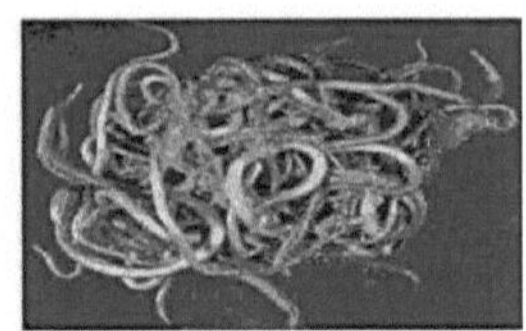

Phylum Mollusca.

Mollusks (Latin **mollis,** soft), constitute a group of animals with bilateral symmetry, with the presence normally of a **head** (except some forms such as pelecypods and scaphopods), a **ventral muscular foot** that can undergo various modifications, to crawl, mine or swim, Its function is mainly locomotion and a **visceral mass** that may be coiled in gastropods and some cephalopods and is covered by the **mantle**, which is responsible for secreting the shell, which may be one, two or eight imbricate plates like the chitons (in some the shell is reduced, in others it is internal like that of the squid or it may be absent).

Zacrisia

The body surface is usually covered by a ciliated monostratified epithelium containing numerous mucous glands and sensory nerve endings. The coelom is reduced to the cavities where the nephridia, gonads and pericardium are located. All types of feeding habits are observed in them; for example: herbivores, carnivores, consumers of suspended particles (filter feeders) and of organic matter in deposit (detritivores) and parasites (ectoparasites and endoparasites).

The digestive system is complete, often acquires a **U-shape**, or may be spirally coiled as in the case of gastropods, there is usually a radula; a strip of teeth always arranged in rows, chitinous constitution and recurved varying in number from 16 to thousands, located on a cartilage base and functioning not only as a scraping organ, but also as a brush, grater, cutter, transporter etc., which assists in feeding, although it has been modified secondarily to accommodate other types of nutrition in various mollusks, the anus opens into the mantle cavity, the **hepatopancreas** constitutes a gland attached to the digestive system and salivary glands are often present.

The blood vascular system is open (except cephalopods) and blood drains from the gills into one or more pairs of atria. From each of these, it passes to the central ventricle, which pumps it through the aorta to distribute it between the sinuses, the blood is generally colorless, the heart is surrounded by a coelomic cavity or pericardial cavity, (scaphopods lack a heart).

The excretory organs are **metanephridia** (one, a pair or two pairs) that drain into the pericardial cavity or not and empty into the mantle cavity. Ventilation is provided mainly by one or more gills (ctenidia), by a **"lung"**, which originates because the edges of the mantle cavity have joined the dorsum of the animal, except for a small opening on the right side called **pneumostoma**, the gills disappear and the roof of the mantle cavity has become highly vascularized. Ventilation is facilitated because the roof of the mantle cavity becomes dome-shaped and there is a flattening of the floor of the mantle cavity (actually the dorsum of the animal). The pneumostoma remains open all the time as a rule or opens and closes with the ventilatory cycle, mollusks can also ventilate through the mantle.

They possess a typical nervous system with pairs of **cerebral**, **pleural**, **pedial** and **visceral** ganglia, linked by longitudinal and transverse connectives and nerves. They possess sense organs of touch, taste, smell, vision (eyespots) or complex chambered eyes like those of cephalopods and statocysts for balance.

Sexes are often separate, some are hermaphrodites, have one or two gonads with ducts, fertilization may be external or internal, mostly oviparous.

In this Phylum there are several classes, of which the best known are:

Gasteropoda or Gastropoda are the most numerous **class** of mollusks. They can be **terrestrial** or **aquatic**, mostly **marine** and **freshwater**.
They are usually **hermaphrodites**, with **cross-fertilization** and mostly with **indirect development through a larva**. Almost all are **herbivorous**, but there are **predators** and even **parasites.**

The **body of gastropods** may have an external shell, like **snails, an** internal shell, like **slugs,** or no shell at all. The **shell** is usually **conical,** with an **operculum, a** closing plate, and spirally coiled; it houses the **visceral mass** and the **paleal cavity,** which in terrestrials functions as a **lung.** The head usually has two **pairs of tentacles**, one **tactile** and the other with **eyes,** and a mouth with a **radula**, its crushing organ, followed by the **digestive** tract and **anus**.

Some gastropods are hermaphrodites, they have both sexes, but they cannot self-fertilize, they need the participation of another individual to reproduce, it is a sexual reproduction. Both individuals lay eggs.

Cuba is home to the polimitas, a species of land snails that are very colorful because they are protected from the siege of collectors. They are highly prized in the eyes of tourists visiting the eastern region where they are endemic.

Class Bivalve or Pelecypoda

Also called **Lamellibranchs** (with lamellar gills) and **Pelecipods** (with axe-shaped feet), they have an external shell formed by two articulated **valves** that cover the body completely. The head is not differentiated. The **foot** is **muscular** and with it, although they are sedentary, they make small movements. The **mantle** is formed by **two lobes** that **segregate the valves** and **enclose the paleal cavity**, where the **lamellar gills**, the **heart** and two **kidneys** are located; many have **siphons** to regulate the flow of water. They feed by filtering microorganisms that are carried to the mouth by water currents.

Most are **unisexual**, with **larval development**. **Fertilization is external or on the female's paleal cavity**. Most are **marine** and a few **freshwater**; they live **fixed to the substrate**, like **mussels**; **buried in sand**, like **clams**, or **inside galleries** they excavate in rock or wood, like **jokes**. (Marine lamellibranch mollusk of vermiform appearance, with disproportionately long siphons and very small shell, which leaves most of the body uncovered. The valves of the shell, functioning as jaws, perforate the submerged wood, practice in them galleries that the animal itself coats with a calcareous material secreted by the mantle, and thus cause serious damage to ship constructions). Similarly, in Cuba, green mussels (Perna viridis) of Asian origin are currently developing and were apparently introduced through the hulls of ships or bilge water from maritime traffic resulting from the accelerated petrochemical development of the Bay of Cienfuegos (unintentional introduction). In 2008, green mussels clogged the cooling channels of the Cienfuegos Thermoelectric Plant, forcing it to stop for several days in order to extract these organisms. On one occasion, 30 trucks were extracted in one day. In these cases they are considered an invasive exotic species.

Ostra

Class Cephalopoda

Their name derives from the Greek ***képhalos,*** "head", and **podos**, "feet". They are **unisexual,** with **internal** fertilization and **direct development**; all **marine and free-living. Carnivorous,** with **highly developed eyes.** They are divided into two groups: **octopods**, with **eight tentacles,** like octopuses, and **decapods,** with **ten tentacles,** like **squids.** Most of the living ones (except nautiluses) **do not have an external shell,** but, except for octopuses**, they have an internal piece of cartilage that protects the cephalic ganglia**, reminiscent of the **vertebrate skull**. **They** have **a muscular** body**,** especially the **arms or tentacles,** which **are the transformed foot** and are provided with very powerful **suction cups**. They have a mouth with a **multi-toothed radula**, a **crushing** stomach and a **digestive tube** ending in a **rectum**; next to it are the ducts of the **ink sac**, a liquid that is expelled into the water through the **siphon** to form a black cloud that hides the animal and **is used for protection.** The **siphon** also serves as a **propulsive organ** when it expels water with force.

Pulpo

In a general sense, mollusks constitute the fundamental basis of food and handicrafts of various cultural groups from different times and places. The mother-of-pearl shells of mother-of-pearl and other species are used to make industrial handicraft objects, which are highly appreciated and of great value. They are also considered of great importance for their participation in the human diet.

Babosa

Colmillito de elefante

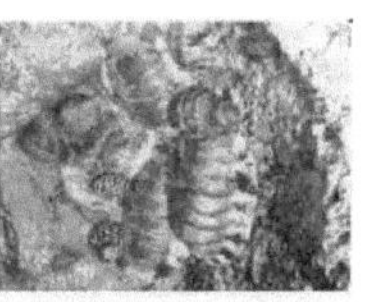

Quitones o cucarachas de mar

Phylum Annelida.

Annelids come from the Latin **annellus,** which means ring. These animals are one of the most important types of non-chordates; approximately more than 15,000 species have been described. They have **bilateral symmetry**, i.e., their body can be **divided into two** symmetrical **parts**, with the **cephalic** (anterior) and **caudal** (posterior) regions more or less differentiated. They have an **elongated, cylindrical** body **covered with a flexible cuticle that grows with the body.** It is divided into **external rings**, which **may or may not be equal**, and into **internal segments or metameres**, separated by

partitions and **always with the same structure**: its **own body cavity filled with liquid that serves as a skeleton, hydroskeleton**, and its own **locomotor, respiratory and excretory organs (Nephridia)** to eliminate waste substances. **Metameres** are **linked by** three common apparatuses:

1. **Digestive**. They have a complete digestive tract with mouth and anus.
2. **Circulatory**. Formed by blood vessels, one **dorsal** and one **ventral**, joined by **lateral vessels in each segment**.
3. **Nervous**. A **double nerve cord runs through the body**, with two **nerve ganglia** in each **metamer**, joined in the form of a rope ladder; in the **cephalic region, two cerebral ganglia** join the cords.

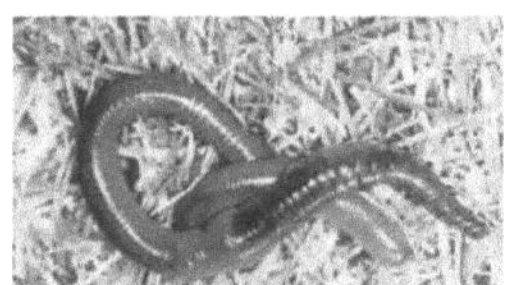

Earthworm

These animals have a body composed of numerous segments, rings, **metameres** or **somites**, their metamerization is generally manifested both in the external and internal morphology, that is, in the muscles, nerves, reproductive organs, blood and excretory vascular system. It should be noted that the segmented part in annelids is fundamentally circumscribed to the throne and that the head represented by the prostomium and the **pygidium** or terminal portion where the anus is housed, do not constitute segments. In these worms, each ring or segment carries a series of short, even chitinous setae or chitinous **setae**, which have the function of increasing the traction to the substrate by the animal when it crawls.

Within the group, earthworms with their long, cylindrical, ringed bodies are the most familiar (oligochaete class). Most earthworms live in moist soil, although they can also live in fresh water and are sometimes found on the sea coast; certain species are parasitic.

They are hermaphrodites. During copulation. Mating in the opposite direction, they press their ventral surfaces together and remain attached by a substance secreted by the clitellum, a thickened ring in the epidermis. They exchange spermatozoa, which are carried posteriorly to the

The coccoelus and are stored in the seminal receptacles of the other worm. They then separate and a few days later the cocoon secretes a membranous cocoon containing a thick fluid. As the cocoon slides down the head of the worm, eggs are deposited in it, which exit through the female pores, and sperm are deposited in it. Once the cocoon has been released, its openings close so that a capsule is formed inside which the fertilized eggs are transformed into tiny worms. This complex reproductive pattern is an adaptation to terrestrial life.

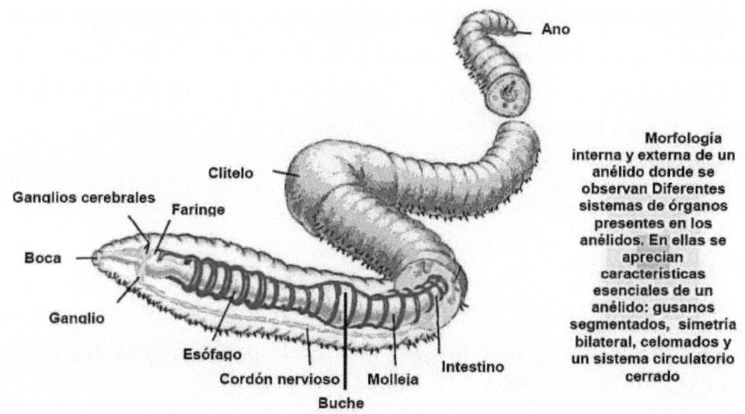

Morfología interna y externa de un anélido donde se observan Diferentes sistemas de órganos presentes en los anélidos. En ellas se aprecian características esenciales de un anélido: gusanos segmentados, simetría bilateral, celomados y un sistema circulatorio cerrado

The polychaetes (Class polychaetes), comprising primarily marine or estuarine worms, plus a small number of species that inhabit fresh and coastal waters and other sapient species inhabiting the supralittoral soil, polychaetes, as their name indicates, have several quetae per somite, starting from each quetal sac, the best known species is undoubtedly the fire worm.

Leeches (Class hirudineos) are mostly freshwater leeches, observed in swamps, lagoons and slow-moving streams in many parts of the world.

Archianellids (Class Archianellidae) are all marine and are a group of small, active, simplified annelids.

The size of annelids is variable, some oligochaetes can measure less than a millimeter in length, however, some giant earthworms, such as ***Rhinodrilus fafneri*** from Ecuador and ***Megascolides australis*** from Australia, reach more than 2 meters in length and 2.5 centimeters in width.

As previously stated, the most accentuated characteristic of annelids is undoubtedly the division of the body into a series of rings or segments, indicated externally by constrictions of the body wall between the segments.

Earthworms are important in nature because they participate in the formation of plant humus, which is why today the artificial breeding of earthworms is practiced in order to use them in the improvement of arable land.

Phylum Annelida.

Arthropoda, derived from the Greek **arthros**, meaning joint, and **podos**, feet, comprise the largest group of species in the living world. More than three-quarters of all known forms of the animal world are included in this group. To date, more than one million species have been described within this group.

Arthropods have adapted to life in the most diverse habitats. They are common on land, in water and in the air, and are often found in places where other animals cannot survive. Undoubtedly, they have achieved great biological success, which has been possible because they have developed structures that have allowed them such extraordinary adaptability.

It is a group formed by approximately 1,100,000 species, which includes **80% of the animals**, of which **insects**, with about one million, **are the most abundant**. They are very ancient invertebrates and the most complex in organization, living everywhere on the planet. Their anatomy and life forms are extraordinarily diverse, but they **share some common characteristics**.

Exoskeleton. They have an **external skeleton** that covers the entire body, consistent but flexible, formed by **chitin**. Sometimes it forms a **carapace**, which is the reason for the frequent phenomenon of **molting**. This skeleton is made up of **articulated parts** and gives arthropods **support and protection** against **predators and desiccation**.

Articulated appendages. The **legs** and other **appendages**, such as **antennae** and **jaws**, **formed by parts** that **articulate with each other,** are mobile; the **number and shape of the limbs is highly variable and depends on the group of arthropods** and **the function** they perform.

Segmented body, with **bilateral symmetry**, generally divided into **three segments,** which in some cases can be joined:

- **Head**. The anterior, with the brain mass. It presents the antennae, the eyes, simple or compound, and the mouth apparatus, provided with mouth with jaws and **maxillae**, to crush the food.
- **Thorax**. The middle one, with important organs, such as the heart. It has articulated legs and wings, if any.
- **Abdomen**. The posterior, with the excretory and reproductive apparatus.

Systems. They have a **complete digestive tract.** The **nervous system** is highly developed and their **brain** shows a more complex structure. The **circulatory** system is open, with a dorsal contractile vessel that functions as a **heart. Breathing is gill in aquatic and**

tracheal in terrestrial. Tracheal respiration is carried out through a system of tubes, called **tracheae, which** communicate with the exterior through orifices and branch throughout the body, carrying oxygen to all parts of the body.

Reproduction. They are mostly individuals of **two sexes and internal fertilization by copulation.** They are **oviparous** and some form **larvae** that **may or may not undergo metamorphosis.**

Arthropods are divided into **four** major **classes**:

1. With chelicerae and without antennae: **Arachnids**.
2. With jaws and antennae: **Myriapods**, **Crustaceans** and **Insects**.

Arachnid Class

Some **arachnids are beneficial** to humans because they **prey on insects**, although some are **parasitic** and can **transmit diseases**, such as **ticks** and **mites,** and others can **cause very dangerous stings,** such as **scorpions** and **some spiders.**

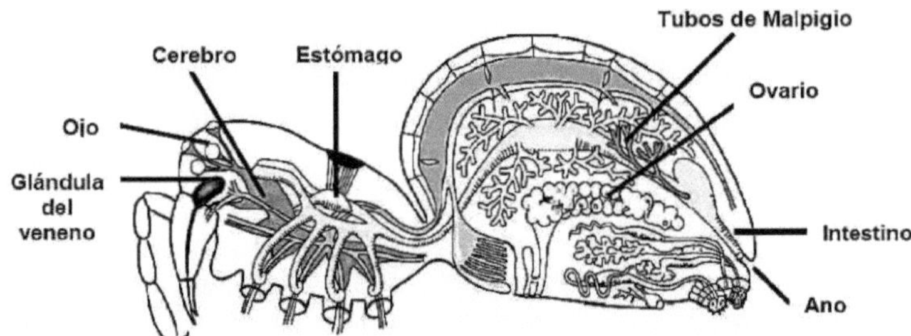

The body with segments not very visible in general, formed by **cephalothorax**, union of **head** and **thorax;** and **abdomen**.

In the cephalothorax are located:

Two **chelicerae**, **nail-like** appendages or **pincers,** which can be **venomous.** They surround the sucking mouth that sucks the soft tissues of their prey, since they have no jaws.

-Two **pedipalps,** sensitive or prehensile organs.

-The **eyes, simple**, almost always **eight**.

-Four pairs of legs for walking. They lack antennae.

Most are **unisexual, internally fertilized, oviparous** and with **direct development**. They are **free-living and predatory**.

Spiders are characterized by chelicerae for injecting venom. They possess silk-producing **spinner glands**, with which they spin webs to capture prey. They have **tracheal respiration.**

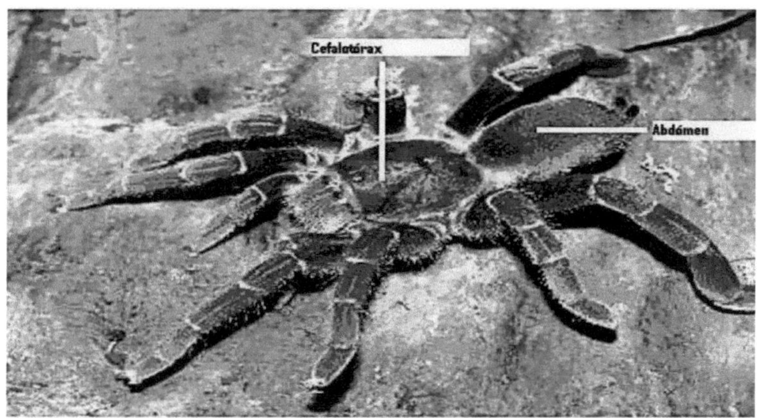

Scorpions are characterized by their **cephalothorax** with **reduced, non-venomous chelicerae**, and **pedipalps** ending **in enormous pincers to hold** and inject **venom** into their prey with the **stinger located at the end of the abdomen.** They have a **comb**, a sensory organ characteristic of these animals. They are **nocturnal and viviparous**, the mothers carry their young on their backs.

Class Miriapoda

There are about 10,000 species, generally **small in size**, although some measure 30 cm.

They have **tracheal respiration** and live in **humid** environments as they do not have an **impermeable cuticle**.

Its body is made up of:

- *J* **Head,** with a pair of antennae and a pair of chewing mouthparts.
- *J* **Trunk** with segments, from 15 to 200 depending on the species, each with at least one pair of legs, hence the names **centipede or millipede**.
- *J.* They feed on decomposing plant matter.

Class Crustacea

For their study they can be divided into **lower** crustaceans, small and without appendages in the abdomen and important in food chains as sustenance for fish, and **upper crustaceans**, with appendages in all segments and of great importance in human food.

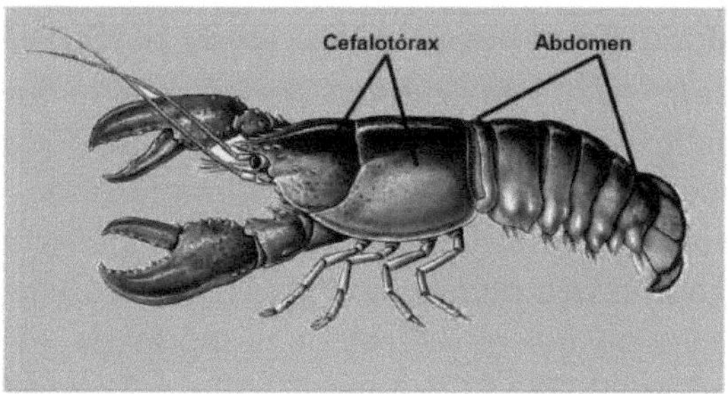

Its body is segmented, covered by a carapace and divided into **cephalothorax** and **abdomen**. In the **abdomen it has appendages,** the latter in the form of blades for swimming.

In the **cephalothorax** they present:

1. Two **pairs of antennae** sensitive to **touch** and **smell;** the first, shorter ones are called **antennules.**
2. **Three pairs** of **chewing** appendages.
3. **Two eyes.**
4. **Five pairs of locomotive legs**, the first one ending in large clamps.

The higher crustaceans have sexual reproduction and separate sexes; in the lower crustaceans there are many hermaphrodites. They are oviparous and have planktonic larvae (in one stage of their development). Most of them are marine and some are freshwater or from humid places, since they breathe through gills. They are carnivorous animals.

Class Insecta

With almost 1,000,000 species, it is the largest group of animals.

They are characterized because their body is formed by:

1. **Head**. With **compound** eyes, a pair of **antennae** with **olfactory** and **tactile** function and **mouth** with appendages, such as **jaws and mandibles**.
2. **Thorax**. With **three segments** carrying the **three pairs of legs and two pairs of** thin, membranous **wings**, although the variety of legs and wings is enormous.
3. **Abdomen**. **Without appendages, it** is formed by **ten or eleven segments,** in which the **orifices of the tracheae** appear.

They are **oviparous**. They can have **direct larval development** (the larva at birth already resembles the adult) or **indirect;** in this case, **metamorphosis** occurs, with **one or more steps from egg to adult.**

The relationship between man and arthropods is very varied. Many are harmful, as they transmit serious diseases to humans and domestic animals, while others are harmful pests for crop plants.

Currently, there is a species commonly known as santanilla *(Wasmannia auropunctata),* which is a very small ant (about 1.5 mm. long), golden brown in color. It is native to Central and South America and has spread to other latitudes. It is reported among the 100 most harmful invasive alien species in the world. Due to its characteristics, it adapts to various habitats, from urban and rural dwellings and their surroundings to cultivated fields, forests and scrublands in general; it creates supercolonies, its activity is around the clock with an opportunistic behavior that allows them to appropriate all the resources available in the environment to the detriment of competing species. Their attack causes intense local irritation that causes discomfort and burning in that area.

But not all arthropods are harmful, as there are many beneficial species; some are used as food, many are involved in the pollination of flowers and others produce substances useful to man.

Due to the enormous diversity of form of this group of animals, it has been necessary to establish a classification that groups them according to their similar characteristics and kinship relationships, in order to facilitate their study.

The diversity of this phylum includes beneficial species such as bees, butterflies, beetles, scorpions, shrimps, lobsters, crabs, etc. because some of them participate in pollination, others their venom is used in the pharmaceutical industry and in others their meat is used for human food. However, species such as cockroaches, mosquitoes, fleas, lice, ticks, etc. are pests and vectors of animal and human diseases that require the improvement of hygienic conditions in general for their better control.

Phylum Echinodermata.

Echinodermata derives from the Greek **echinos**, meaning hedgehogs and **derma,** skin). It consists of about 6,000 species, among which are some of the most common marine animals. They include sea stars (class asteroidea), ophiuroids (ophiuroidea), sea urchins and sea dollars (echinoidea), sea lilies (crinoidea) and sea cucumbers (holothuroidea), as well as several extinct classes.

Starfish

All are radially symmetrical in the adult state and most have a calcareous endoskeleton with external spines. They live on the coast and on the seabed, from the tide line to more than 3,600m depth, mostly free-living and slow-moving; a few are pelagic, but none are parasitic. Some are abundant, but none are colonial; some sea lilies are always stationary and many others are free-swimming; some echinoderms are used by man as food and their eggs have been used in numerous experiments. Starfish can damage oyster or clam beds.

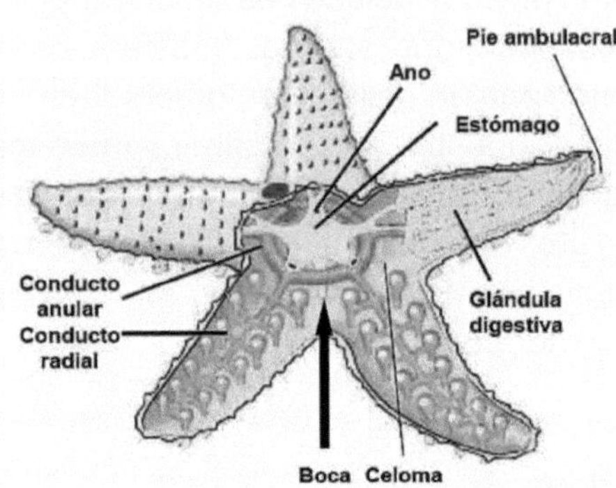

Echinoderms have ordinary radial symmetry, pentaradiate in adults, bilateral in larvae, most of their organs are ciliated, the body surface is formed by five symmetrical radial areas and ambulacrum, in which the tubular feet emerge and alternating between them there are five interambulacrum (inter radii).

The body of these animals is covered by a delicate epidermis and has a resistant mesodermal endoskeleton, formed by movable or fixed calcareous plates, usually with a defined arrangement; often with spines (leathery skin and generally microscopic plates in Holothuroids).

It presents different organ systems. The digestive tract is simple and generally complete (some lack anus).The circulatory system is radial (hemal), but reduced and the coelom is lined by a ciliated peritoneum, usually large, with free amoebocytes in its fluid; part of the larval coelom becomes an aqueous vascular system, which usually has numerous tube feet for locomotion, food capture and respiration, which is verified by minute gills or protractile papules in the coelom, by tube feet and, in holothurians, by cloacal respiratory trees. The nervous system of echinoderms consists of a circumoral ring and radial nerves. The sexes are separate (exceptions are rare), externally equal, the gonads large, with simple ducts; abundant eggs, usually fertilized in the sea, the larvae are microscopic, ciliated, transparent and usually swimming, with remarkable metamorphosis.

There are few viviparous species, some are also produced asexually by division, and many regenerate lost parts rapidly.

Up to this point we have studied different groups of animal organisms that have varied adaptations to the different conditions in which they live, but have the common characteristic that they lack a vertebral column and are therefore called invertebrates.

The following is a study of the animals that are known as vertebrates because they have a vertebral column, and you will see that in comparison with the previous ones, they present a higher level of evolutionary development.

PHYLUM CHORDATA

To this group of animals belong all species that during their embryonic development present the following characteristics:

--Presence of notochord.

--Presence of dorsal nerve cord.

--Presence of pharyngeal or gill clefts.

These, in their adult state, may be modified or replaced by other organs.

For example, in the subphylum Urochordata, where the ascidia, known by fishermen as sea syringe, is found, the three are only in the larval stage, in the subphylum Cephalochordata, where the amphioxus is found, they are there throughout life, and in the subphylum Vertebrata, the notochord is replaced by the vertebral column.

Vertebrates are distinguished by the presence of the vertebral column which constitutes the skeletal axis of their body. This support develops around the notochord and reinforces or replaces it. The vertebral column are bony or cartilaginous segments called vertebrae. In the anterior position is the skull, which surrounds and protects the encephalon, the anterior end of the nerve cord.

Skull and vertebral column are part of the endoskeleton. In contrast to invertebrates that possess an exoskeleton, it is a tissue that grows with the animal and is not replaced by molting.

Another important characteristic of vertebrates is their cephalization, that is, the concentration of nerve cells and sensory organs in a well-defined head.

All vertebrates share other characteristics that, although not exclusive to them, are more developed. They possess a closed circulatory system with a heart endowed with two, three or four chambers; paired kidneys; complete digestive tract and digestive glands (liver and pancreas); muscles attached to the skeleton to guarantee their movement; a nervous system with an autonomous division that regulates the involuntary functions of the internal organs; well-developed sense organs (eyes, ears, smell and taste); two pairs of appendages and separate sexes.

The aforementioned characteristics clearly show the high evolutionary development reached by vertebrates in the kingdom Animalia.

Some of the classes belonging to this phylum will be detailed below.

Class Chondrichthyes.

The class **Chondrichthyes** groups a great variety of vertebrate animals characterized by an internal skeleton of cartilaginous consistency, hence the name of the group, which comes from the Greek **chondros:** cartilage, and **ichthyes:** fish. This class includes, among others, the great diversity of sharks, rays, rays, mantas, bishops and chimaeras that inhabit mostly the waters of seas and oceans around the world. These fish are generally of medium size, ranging from 90 **cm** to 2.5 **m**, but there are species of extraordinary size such as the ladyfish, which reaches more than 15 **m** and is considered one of the largest living vertebrates known; however, some species of sharks and rays do not exceed 30 **cm.** Many representatives of this group are known for their swimming speed, as well as for their feeding habits. Some of them are feared by fishermen and bathers.

All chondrichthyans have one pair of jaws and two pairs of fins. The skin contains placoid scales. Each of these scales is a tooth-like structure consisting of an outer layer of enamel and an inner layer of dentin. The lining of the mouth has larger scales that function as teeth. The teeth of these organisms are immersed in the flesh, not fixed to the jaws; new teeth continuously develop behind the functional ones, and move forward to replace those that are lost.

Shark

Cartilaginous fish have five to seven pairs of gills. A stream of water enters the mouth and passes through the gills to exit through the gill slits so there is a constant supply of oxygen.

They have a digestive tract consisting of an oral cavity, followed by a long pharynx that leads to the stomach and from there to a short, straight intestine to which secretions from the liver and pancreas reach. It ends in a cloaca which receives digestive wastes, urine and in the female the semen of the male. This is characteristic of many vertebrates and opens in the lower part of the body.

They have a complex encephalon and the medulla protected by the cartilaginous structures of their vertebral column. They have sensory organs that allow them to locate their prey by smell and the perception of vibrations in the water, as well as by sight.

Characteristic of all fish is the existence of a lateral line forming a groove along the entire length of the body with small openings to the outside where sensory cells in the channels react to the movement of water.

The sexes are separated and fertilization is internal. Some specimens such as manta rays and some species of sharks lay their eggs (oviparous), while many species of sharks incubate their eggs inside themselves and are therefore considered ovoviviparous. A few are viviparous.

Most sharks are predators, devouring other fish, crustaceans and mollusks.

It is known that some species of chondrichthyans cause harm to humans, but the reality is that man has also caused damage, so many species are in danger of extinction because sharks and rays are a source of food for man and the skin is used in the footwear industry to make bags and shoes. The oil from its liver is widely used in medicine because it is an important source of Vitamin A.

It is therefore necessary, in the different regions of the planet where these species are used for such purposes, to better manage these resources to avoid the loss of these members of the biological diversity.

Class Osteichthyes.

Bony fishes, the name given to this group of vertebrates from the Greek **osteo:** bones and **ichthyes:** fish, are the animals that predominate in the aquatic habitat.

This group is made up of some 20,000 species with the most varied characteristics in terms of habitat, body shape, color and size. Among the diversity are the creole snapper, claria, trout, tench, sardine, sardine, yellowtail, grunt, seahorse, moray eel, bunting, etc.

The abundance of bony fishes is considerable; they live in both salt water and fresh water, have colonized all possible biotopes for their life and have successfully exploited them, in this influences the resistance to unfavorable environmental conditions, even through dietary modifications of singular importance. Most of them are piscivorous, but others ingest everything from plankton to all types of aquatic vegetation. Therefore, it is not surprising that bony fishes outnumber cartilaginous fishes by almost 8 times.

Bony fishes are basically characterized by having the body covered with scapes of dermal origin. Most species have both medial fins and even fins, with cartilage or bone rays. Unlike chondrichthyans, they have a lateral protective fragment of their body wall called the operculum, located at the back of the head and covering the gills. Thanks to the movement of the operculum, there is a circulation of water that enters through the mouth and exits through it, allowing bony fish to obtain the oxygen dissolved in water, which is necessary for respiration.

An aspect that also distinguishes them from chondrichthyans is the presence of a swim bladder, a hydrostatic organ that allows the fish to change the density of its body and, in this way, to ascend, descend or remain stationary at a certain depth. Cartilaginous fishes do not have a swim bladder and must swim constantly, otherwise they sink.

Bony fishes are usually oviparous. They are fertilized externally and lay a large number of eggs from which fry hatch. It often happens that these large numbers of eggs and young fish serve as food for other fish. Many fish species build nests and take care of them until the young hatch.

The **creole snapper** is a species that generally inhabits the waters of our insular shelf, where it usually moves in large groups, its size ranges between 40 and 80 **cm.** in its adult state, and is easily distinguished by the bright colors ranging from red or pink in the ventral region of the body and fins, to the greenish color of the head and dorsal region, where a large spiny fin is observed.

This shade of red and pink allows it to go unnoticed in its habitat, so that other fish that feed on it cannot easily discover it, so its color is a form of masking for the snapper. This species is one of the exportable economic items due to the quality of its meat. Along with snapper, there is another diversity of bony fishes that represent economic items for export as well as for national consumption, since the Cuban platform, due to its insular character, the existing marine ecosystems and the barriers of the sea, is an important economic item.

coral reefs allow the development of a diversity of species that are widely commercialized.

Some freshwater species have characterized the Cuban fauna of bony fish such as the biajaca criolla, the manjuarí (in danger of extinction), the trout, etc. For years they have served as food for the peasants and other part of the population in general. Currently, the introduction of some species such as claria, tench and tilapia for economic purposes has favored this food line.

In spite of this, the claria, also known as catfish *(Claria gariepinus)* of African origin, was intentionally introduced by the Ministry of Fisheries for food purposes due to the nutritional characteristics of its meat. It is a predatory omnivore that has become surprisingly widespread throughout the country, under various circumstances and environmental conditions. In the study of the stomach contents of the clarias, remains of amphibians, reptiles, mollusks, birds, mammals and vegetation have been found, which demonstrates all the feeding routes it possesses. Its vertiginous development has caused them to attack the Creole biajaca, which is almost disappearing from Cuban rivers and lagoons, and has reproduced so vertiginously that it is practically at this time a species of intelligent management and control because it is considered, under these circumstances, an invasive exotic species.

Another invasive alien species is the lionfish *(Pterois volitans).* Its presence in Cuba was reported for the first time in 2007 in the eastern region of the country. Since then, its dispersion has been rapid and it is found in almost all our coasts.

It is a species of direct predation, with a high competitive level in its environment and overpopulation due to its easy reproduction. The venom of its spines can cause serious damage to humans. To date, several cases have been reported of people intoxicated with the venom of this fish.

Amphibia Class.

Amphibians are the oldest extant representatives of terrestrial vertebrates. The name of this class comes from the Greek words **amphi**, meaning both or double, and **bios**, life. That is, the animals that make up this group generally have a "double life" since most of them develop one stage of their life cycle in water and the other on land.

At present, approximately 2,500 species of amphibians are known. Of these, frogs and toads are well known to us because of their extraordinary abundance in ponds and humid places in the country.

Amphibians **are classified into three orders**: **Urodela**, consisting of salamanders, newts and necturans, all with long tails; **Anura**, which includes toads and frogs, with no tails and specialized jointed legs for jumping; and **Apoda**, of the caecilians, which are apodous and wormlike.

Observe some representatives of these orders

Apoda	**Urodela**	**Anura**
Cecilia	Salamandra	Rana

These animals are characterized by having four limbs (although some have two or simply no limbs) with or without interdigital membranes. See the following image:

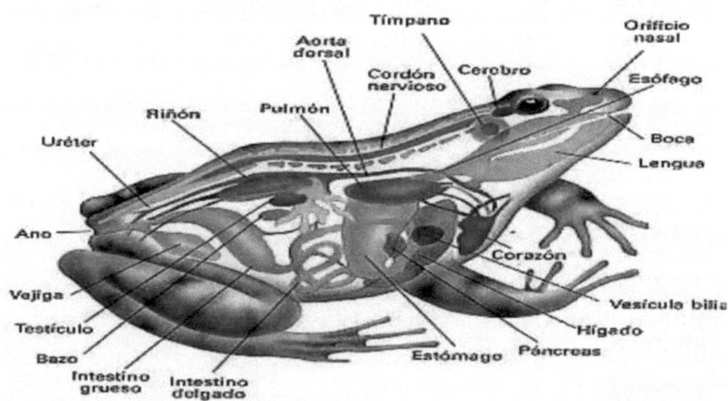

The skull has two occipital condyles, a pelvic girdle attached to the vertebrae and double circulation with mixing of blood in the last ventricle. The amphibian heart has two atria that receive blood and a single ventricle that pumps it into the arteries. A double circuit of blood vessels keeps the oxygen-rich blood partly separated from the deoxygenated blood. The blood passes through the circulation process to the tissues and organs of the body. Then, after returning to the heart, it is directed through the pulmonary circulation to the lungs and skin, where it is re-oxygenated. The oxygen-rich blood returns back to the heart to be pumped and distributed throughout the body.

Pulmonary respiration is also reinforced by cutaneous respiration, because their skin is moist (due to the presence of many mucous glands) and vascularized, and by the buccal mucosa. In larvae it is gill-like. Adult amphibians do not depend only on primitive lungs for respiratory gas exchange since, as mentioned above, their skin also functions as a respiratory organ because it is moist and has glands that keep it moist, in addition to being well supplied with blood vessels.

The presence of such glands in the skin to keep it moist and prevent desiccation is another important adaptation to the conditions of terrestrial life.

The eyes have eyelids which are an adaptation to terrestrial life to avoid the action of solar radiation and thus remain moist.

The lateral body line is present only in larvae and in some adults that develop their entire life in the water.

Amphibians have 10 pairs of cranial nerves, excretion is through the mesonephric kidneys and through the skin, and reproduction is oviparous.

An important aspect in reproduction is that despite having invaded the land where some adult amphibians are very successful as terrestrial animals and can live in quite dry places, most of them are dependent on water to perpetuate their species, because their egg lacks a resistant envelope to avoid the effects of solar radiation. For this reason they go through a series of transformations known as metamorphosis.

Eggs and sperm are usually deposited in water in a gelatinous substance that holds them together, frog and toad embryos develop into larvae called tadpoles that have tails and gills but should not be confused with some type of small fish. These feed on aquatic plants and after a while begin to undergo a series of transformations (metamorphosis).

The gills and gill slits disappear, the tail is gradually shortened, the front legs develop, the digestive tract is shortened and the diet changes from the small plants on which they used to feed to a carnivorous diet, the mouth widens and a tongue with special characteristics develops, its base is in the front part of the mouth and the back part is rolled up to be able to launch it to hunt prey, the tympanic membrane and eyelids appear.

Some species of salamanders do not undergo complete metamorphosis, retaining many larval characteristics in their adult phase and thus becoming sexually mature without having completed metamorphosis (this is known as neoteny). Observe in the figure the process of metamorphosis.

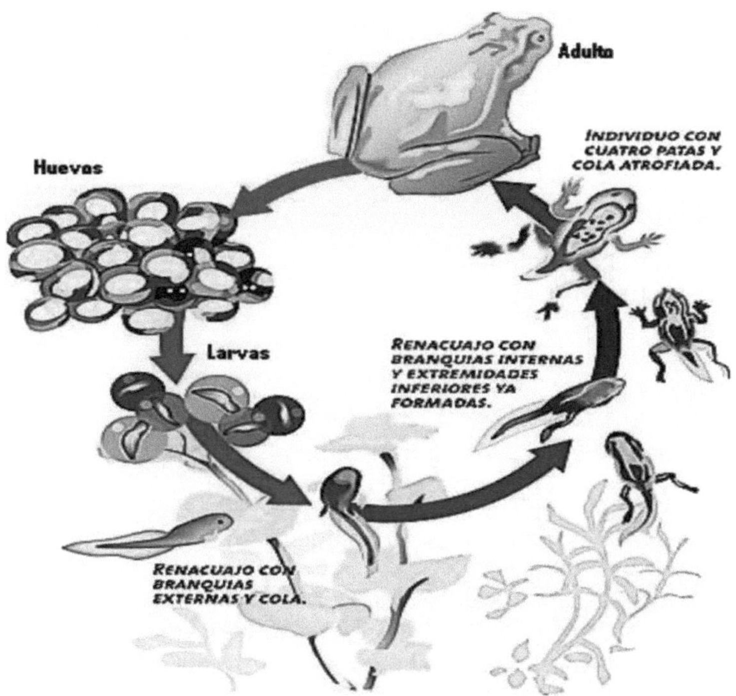

The aspects described so far are sufficient to understand that amphibians, from an evolutionary point of view, are superior to fish because they have invaded the land, but they cannot conquer it since they still have anatomical and functional characteristics that depend on water, otherwise, adaptations to counteract the effects of terrestrial life.

Amphibians are very important for the life of man and the rest of the organisms in nature. Being carnivores, they feed on insects such as flies, mosquitoes, small cockroaches and spiders, among others, so they act indirectly as biological controllers, since these insects can cause diseases in humans and other animals.

Amphibians have the ability to change color to adapt to the environment where they live and to blend in so as not to be attacked by other organisms, this is called mimicry. The color change depends on hormonal processes regulated by the pituitary gland. Some of these secreted substances are toxic.

Some species, such as the bullfrog, have a healthy and highly valued meat for food and are also used as a source of export.

Class Reptilia.

Reptiles are the first vertebrates to be fully adapted to the terrestrial habitat, so they are abundant in dense forests, coasts, marshes, high mountains, etcetera.

At present, approximately 6,300 species of reptiles are known, including such well-known animals as lizards, snakes, sea turtles, land turtles and crocodiles, among others.

The name of the class comes from the Latin **repere: to** crawl (to creep) and refers to the form of locomotion common to most of these animals.

Reptiles are truly terrestrial organisms because they do not need to return to water to reproduce. Reptiles are said to be the first organisms to conquer land. Amphibians had already invaded the land, but they depended on water. We will now analyze the characteristics of reptiles that allowed them to adapt to life on land.

Their bodies are covered by hard, dry, horny scales, which protect them from desiccation and predators, but they cannot use them for gas exchange as amphibians did, so their lungs are better adapted because they have chambers and a larger surface area for gas exchange.

Iguana

Most reptiles have a heart divided into three chambers and the ventricle has an internal but incomplete septum or division that facilitates the partial separation of oxygen-rich and oxygen-poor blood, which favors the oxygenation of all the tissues of the reptiles' body. Crocodiles do have four cavities in the heart, which allows purified blood not to mix with impure blood.

Another adaptation to terrestrial life is the way in which reptiles eliminate waste substances from their bodies without discarding much water, since in the process of urine formation, much water is absorbed and not expelled to the exterior.

The body temperature of reptiles cannot be controlled by themselves, as is also the case in fish and amphibians, but depends on environmental conditions. However, reptiles have mechanisms to raise their body temperature. Sometimes you have observed lizards exposed to the sun. All this influences their metabolic processes. A reptile exposed to the sun increases its metabolic capacity and therefore has greater activity, consequently reptiles look for mechanisms to increase their body temperature with respect to that of the environment. All this will allow you to conclude that reptiles manifest greater activity and therefore better performance in warm climates than in cold ones.

Most reptiles are carnivores, although some, such as turtles and lizards, are herbivores.

Its forelimbs and hind limbs generally have five toes and are perfectly adapted for running and climbing.

Their sensory organs allow them to locate prey.

One of the adaptations to the terrestrial widow of reptiles is in their way of reproduction. It has been previously stated that they do not need water to reproduce. Although some species such as turtles and crocodiles (including alligators) live in water, during the breeding season they come out onto land to lay their eggs until they produce their offspring.

The reptile egg is covered by a leathery protective shell which helps prevent the embryo inside from being damaged by external conditions of terrestrial life, such as high temperatures, the action of other animals, etc.

Before the formation of this shell, fertilization in reptiles occurs inside the female after copulation where the male, through his copulatory organs, deposits the spermatozoa inside the female.

After fertilization and under the protection of the egg shell, the embryo develops in a liquid chamber surrounded by a membrane called **amnion**. Under these conditions the embryo is kept moist and the liquid chamber acts as a shock absorber against external effects, i.e. possible mechanical shocks or blows.

Other envelopes within the egg contribute to the successful development of the embryo. The **chorion** is found surrounding the embryo and the rest of the extraembryonic membranes, so its fundamental function is protective. The **allantois** forms a third pocket through which the embryo takes the necessary oxygen that arrives through the porous egg shell. The allantois also functions as a reservoir for the embryo's excretion products.

Once the embryo's development time has elapsed, the fully formed animal breaks the shell that surrounds it, emerging a small specimen with similar characteristics to its parents, which quickly adapts to the new living conditions.

There is a great diversity of reptile species, from majaes, terrestrial and aquatic turtles, crocodiles and alligators to lizards of different shapes and sizes. This is why they are **grouped into three orders**: **Chelonia**, which includes turtles ranging in size from 8 cm. to 2 m. and their bodies are covered by a protective shell made up of bony plates with overlapping horny scales forming a carapace that allows the organisms to live inside it and can protect the head and legs from external effects.

Squamata consisting of lizards, snakes and iguanas are the most modern reptiles and their skin is covered by rows of scales that overlap like the tiles of a roof forming a continuous armor that can undergo molting as the individual grows. Some, such as lizards, have the ability to regenerate the tail when it has been damaged.

Crocodilia include crocodiles and alligators among others. Almost all of them live in swamps, in rivers or on sea shores, where they burrow in the mud and feed on various types of animals. Crocodiles are the largest living reptiles and can measure up to 7 meters.

The image below shows some specimens of the three orders described above.

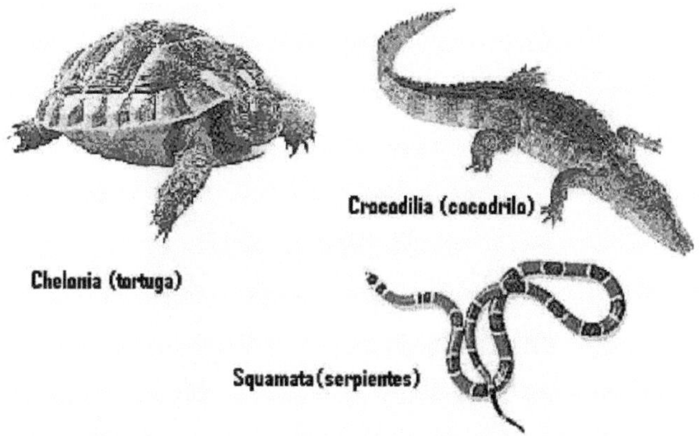

Reptiles are widely represented in Cuba. These animals are of great importance as **biological controllers in nature.** Thus, for example, the maca of Santa Maria contributes to the control of mice and rats in the sugar cane fields. On the other hand, lizards and snakes feed on insects harmful to crop plants.

It is also known the value of serums prepared in other countries from snake venom, which have considerably reduced the mortality rates caused by snake bites.

The importance of reptiles to human life has long been known. For example, the skin of crocodiles has traditionally been used for making shoes, handbags, etc.

For these reasons, the Cuban State has taken measures to ensure the conservation of this reptile species in our country. Some places in the Zapata swamp have been set up as crocodile breeding and protection centers.

Class Birds.

The term birds comes from the Greek word ***ornis,*** *which* is one of the most developed groups of vertebrates in the animal kingdom. These interesting animals, which comprise more than 8,600 species, are easily recognized by humans because their bodies are covered with feathers, flexible structures that are very strong for their light weight. They protect the body, reduce water loss through the body surface, reduce body heat loss and participate in flight by presenting the air with a supporting surface.

Forelimbs in the form of wings adapted for flight in most species and hind limbs for walking or swimming.

They have a horny beak. Beaks and legs are modified according to the places where they live and the way they feed.

All these characteristics distinguish birds from other vertebrates.

In addition to wings and feathers, birds have other adaptations for flight. Their bodies are compact and aerodynamic. Their bones are strong, hollow with large air spaces, making them very light, many of them fused together, which gives them the rigidity necessary for flight.

Because of their ability to fly, most of them have been able to invade numerous habitats, many of them unreachable to other animals, which gives them certain advantages in fleeing from their enemies.

The jaw of birds is light and instead of teeth they have a horny beak. The lungs are very efficient because they have very thin wall extensions called air sacs, which occupy free spaces between the internal organs and inside some bones. They have a heart divided into four chambers and a double circulation where the blood supplies oxygen to the tissues and then is oxygenated again in the lungs to be pumped once more to all parts of the body. Their blood is warm and they maintain a constant body temperature so they can live in cold climates without difficulty.

Bird excretions are produced through the cloaca because they lack a urinary bladder. This allows them to maintain a reduced body weight, which is also important for their ability to fly.

The metabolism of birds is very high to obtain the energy necessary for flight, however there are different ways of acquiring food through their beaks. Some feed on small seeds or fruits, worms, mollusks. Some hunt rodents, rabbits, snakes, etc. with their clawed feet. These foods are deposited in the crop, a peculiar organ of birds. They also have a muscular, thick-walled gizzard, where food is reduced to small portions to facilitate digestion. Many birds ingest small pebbles that help in the gizzard to degrade the food.

Birds are characterized by a well-developed nervous system with a brain larger than that of reptiles. They rely heavily on their sense of sight, their eyes are relatively larger than those of other vertebrates. Hearing is also well developed.

In spite of the peculiar shape of the birds' bodies, they show a great diversity in size, plumage coloration and the different shapes of their beaks and legs. Thus we can highlight the small size of a canary next to the large size of the condor, the contrast of the plumage of the negrito with the wonderful colorful feathers of the male peacock, and the extravagant beaks of the pelican, the parrot, the woodpecker and the zunzún, among others.

Many birds such as the nightingale and the mockingbird possess the faculty of song, one of the most beautiful manifestations of these animals, which has attracted the attention and admiration of man since ancient times.

This group of animals has also represented a valuable source of food for mankind due to their meat and eggs.

Birds are characterized by their movement to other places (migrations), the way they make their nests and their song, aspects that distinguish them from other animals.

One of the most notable seasonal phenomena of birds is the habit of many species to move, generally from northern to southern regions, in search of more favorable conditions, such as better temperatures and food, among others.

During migrations, birds generally use fixed routes during certain times of the year.

These animals do not learn the migratory route from their parents, as the young regularly leave first.

Some scientists consider it probable that the ability of birds to follow a fixed course depends, in part, on the ability to orient themselves by observing the position of the Sun or the stars; but this is only a hypothesis, since it is not yet known as a certainty.

As an example of migratory birds we have the **Florida duck,** which breeds in North America and in the winter migrates south where it is common to find it in our lagoons and other shallow freshwater places; another example is the **bijirita**, which is a common winter visitor in our country.

Penguins, strange flightless birds with modified forelimbs in the form of flippers, perform northward swimming migrations.

Another temporary phenomenon in the life of birds is **nesting**. Most birds build their nests on the ground, on branches, on rocks, or other places, and there they hide their eggs or young from their enemies.

The construction of the nest is characteristic in the different species, usually the male is in charge of collecting the material and the female builds the nest, but this may vary. During the breeding period, the male generally defends his territory and does not allow other males or enemies to enter.

Birds are the most perfect nest builders, due to the need to protect their young from enemies and weather variations.

Some birds, such as the well-known **eagles,** build their nest and remain in pairs throughout their lives.

The ability to return to the nest is a remarkable phenomenon in birds, apart from migration.

Some species of trained pigeons return to the nest when released at distances of more than 800 km. It has also been demonstrated that they can return to their point of origin without prior training. This ability of some birds to return to the nest probably responds to the fact that they fly to their point of origin whenever they can see the sun.

The number of eggs that the female lays in the nest varies according to the species, but the number of birds that build their nests in safe places is always lower than those that nest on the ground. For example, the **quail** that nests on the ground, exposed to attack by enemies, lays up to fifteen eggs, while the **common swamphen**, which builds its nest hidden in the branches of trees, lays only two eggs.

The color of eggs in birds can also vary; for example, shorebirds nesting in places with little protection lay eggs with spots of coloration very similar to the surrounding substrate, while **owls,** which build their nests in cavities, where they are protected from their enemies, lay eggs with white shells.

Egg incubation is also variable. The birds incubate for a period of 12 days, the pheasant between 21 and 26 days and the ostrich incubates the eggs from 42 to 60 days.

During the incubation period in birds, the developing embryo requires a high temperature, which is usually provided by the female laying on the eggs.

Singing is one of the most beautiful faculties that some birds possess. It constitutes one of the factors of the nuptial exhibition and allows the recognition between the male and the female to perform the copulation. It also constitutes in some species a threat between males, other times it is evidence of certain hierarchies between parents and offspring and sometimes it is considered a means of warning in case of danger.

Some species, such as the **nightingale,** emit beautiful notes all year round, but other birds, such as the **mockingbird** and the **fermina**, tend to do so only at certain times of the year. In such cases, the song may be associated with reproduction or migrations.

About 380 species and subspecies of birds have been recorded in our country, of which 28 are endemic to our archipelago. Among the endemic birds reported so far and that stand out for their beautiful plumage, we have the **partridge pigeon** and the **tocororo**, our national bird. No less interesting are some birds that are true rarities due to their difficult location, such is the case of the **zunzuncito,** which is found only in some parts of the country.

A good number of birds, on the other hand, are very common and have become associated with man, so much so that sometimes they even share their premises, as is the case with the **totíes** and the **mayitos**, which gather in large flocks and go to sleep in the trees of the city parks.

Other birds, such as the **catey** and the **cao,** bring harmony to the fields where they live, due to their characteristic voices. Although most of our birds are diurnal, that is, they carry out their activities during the daytime, other birds that begin their activities in the evening, such as the **owl,** the **querequeté,** the **siguapa and the sijú cotunto,** are still interesting.

Since ancient times, man has used birds for food and ornamentation, which has meant that over the centuries this group has achieved great importance in nature and in the economy.

Many of them, such as pigeons, chickens, pheasants, among others, provide man with an invaluable source of food.

In our country, the development of poultry farming has reached a great boom after the triumph of the Revolution and it is known from the Report of the Central Committee of the PCC to the First Congress that the production of eggs rose to seven hundred million, which represents six times that obtained in 1958.

Poultry meat also had a marked increase on that date; all of which shows that our country occupies a prominent place in poultry farming worldwide.

Among the most important poultry breeds exploited in our country we have the Cornish, which is a meat producer and is characterized by the great development of the pectoral muscles (breast), and the Leghorn, a great producer of small eggs that are rarely laid.

But not only birds are important as a source of food, many of them consume harmful insects and seeds of plants that are weeds in certain crops. Owls are predators of rats and mice, which cause great damage to crops and to humans in general. However, some birds are harmful to man, such is the case of the **chambered owls,** which invade rice fields to feed. Others cause some damage, but not of consideration, to the fruits from which they take their juices or seeds, like the **parrots,** the **cateyes,** the **woodpeckers,** etc.

Due to the disappearance of many of our forests carried out by man in order to use those lands to cultivate the land, the birds of our country have been remarkably restricted.

Some birds, such as the **real woodpecker,** are found today only in certain areas of the eastern provinces, others have been ruthlessly persecuted by man, such is the case of the parrot, sparrow hawks and some species of pigeons that are in danger of extinction. For this reason, it is necessary to protect birds and create a conscience based on the love of nature, which will allow us to contribute positively to the work being done by our State in the task of reforestation, which will eventually create the necessary conditions to preserve our birds.

The birds' ability to fly made it easier for them to conquer new territories. Different species have arrived and established themselves naturally in our country, such is the case of the Cattle Egret, the Loggerhead, the Bahama Duck, the Cowbird, and more recently, the Nun and the Turtle Dove. Some species, such as the Cowbird *(Molothrus bonariensis),* could have a negative impact on the native avifauna. This bird is one of the most distinctive problems in our ornithological fauna. Its damage is based on nesting competition with other species, affecting their reproduction and behavior, since it steals the nests of many of them, such as the tocororo, the mockingbird, the royal thrush, vireos and others.

Class Mammalia.

The term **Mammalia (**from the Latin **mamma**: breast), which identifies these animals, was assigned by Linnaeus and responds to one of their main characteristics, i.e. the presence of mammary glands that have reached a greater complexity in their structure and vital processes, and therefore constitute the most evolved group in the animal world. Zoologists have described about 3,700 species and several thousand subspecies of mammals, which illustrate the great diversity of this group of vertebrates.

Mammals are distinguished from the animals previously studied by their hair-covered bodies, milk-producing mammary glands to feed their young, and the differentiation of their teeth into incisors, canines, premolars and molars.

They possess a muscular diaphragm that considerably aids the respiratory processes of inspiration and expiration for pulmonary ventilation. The nervous system is more developed than any of the previously studied classes. It is characterized by a large and highly complex brain with an efficiently developed cerebral cortex.

Like birds, mammals are endoderms: they maintain a constant body temperature, determined by the presence of hairs on the skin that act as insulation; the heart has four cavities and a well-delimited double circulation and the presence of sweat glands. Red blood cells, without nucleus, are excellent oxygen transporters.

Fertilization is always internal, and except for some species such as monotremes, which lay eggs, mammals are viviparous. Another important characteristic is that most of the representatives of this class form a placenta, an organ that allows the exchange between the embryo and the mother, through which it is nourished and eliminates waste.

Mammalian limbs are adapted in various ways: for walking, running, climbing, swimming, digging or flying. In mammals that walk on all fours, the limbs are more directly under the body than in reptiles, which contributes to speed and agility.

Due to their evolutionary advancement, they are biologically successful because they adapt easily to new situations; they stand out for their ability to find food and shelter and to care for their young. In this way they have reached a very wide distribution on land and their habitats are very varied. They are frequent on land, under the surface of the ground, in rivers, lakes, seas and oceans, in deserts, glaciers and even in the air. They are rightly animals widely adapted to the most diverse environmental conditions.

On the other hand, mammals also show an extraordinary diversity of shape and size.

The shape of a horse, a seal, a hutia, a monkey, and the size of a rat, a dog and a whale are very different.

Mammals have always been a group of extraordinary interest, since man has always used them for his own benefit. They were indispensable in the life of primitive man and even today he continues to make use of them, obtaining meat, skins, wool, milk and many other products.

The diversity of existing mammals has forced us to classify them as follows: The so-called **monotremes**, which are oviparous mammals, that is, they lay eggs, such as the platypus and the echidna. Their eggs can be carried in an abdominal pouch or incubated in the nest. **Marsupials**, which include pouched mammals such as kangaroos and opossums. The embryos begin their development in the maternal uterus, after a few weeks, still at a very undeveloped stage, the young are born and move to the marsupium, where they complete their development due to the fact that in the marsupial pouches there is the nipple with a mammary gland to which they cling. **Placental** mammals are characterized by the existence of a placenta where the blood vessels of the embryo enter in close relationship with those of the mother, which makes possible the exchange of materials. It does not usually happen that the bloodstream of the mother and the embryo mix, this exchange occurs through the structure of the placenta. Thanks to the existence of the placenta, the embryo remains inside the mother's uterus until its birth.

Placental animals are very diverse in their forms and adaptations to the environment, for example there are the best known, the carnivores, where we can mention cats, dogs, seals, walruses, sea lions, otters, etc. Others such as squirrels and rats are rodents because they gnaw food with their teeth. Rabbits, hares and pikes have long legs adapted for jumping and many have long ears. Elephants, on the other hand, have long muscular and very flexible trunks with thick skin and large size as adults. Other mammals such as whales, dolphins and porpoises have adapted to their aquatic way of life. They are fish-shaped because they have fin-like limbs and no hind limbs, but all of them breathe through lungs and feed their young through scales, their skin has no scales and their bodies are covered with hair. There are others, such as the manatees, which are herbivorous mammals that also have the characteristics of the previous ones. Cattle, sheep, pigs, deer, giraffes are herbivorous mammals with hooves on their feet, most have two digits, but some have four, most of them are ruminants, that is, they deposit the ingested food and then rest and swallow the food to chew it (rumination). Once taken to the stomach they are transformed under the action of certain bacteria that decompose cellulose. Bats are mammals adapted to flight; a fold of skin extends from the elongated toes to the trunk and hind legs, forming a wing. Their flight is controlled by a kind of biological sonar: they emit high-frequency screeches and are guided by echoes reflected off obstacles. They feed on insects and fruits or suck the blood of other animals. They often transmit diseases such as yellow fever, paralytic rabies and leptospirosis.

Primates, which include humans along with others such as monkeys, apes and lemurs, are characterized by highly developed eyes and brains, nails instead of claws, opposable thumbs and forward-facing eyes.

In Cuba there is a diversity of mammals such as dogs, cats, rams, horses, cows, pigs, hutias, rats, goats, etc. and on our coasts we find the Antillean manatee and dolphins.

However, some of these animals were introduced by humans as pets (dogs and cats), for work (mules), for hunting (wild boar, deer) or as pest controllers (mongoose). Many of these animals currently have wild populations and are generally harmful to native fauna. Some examples illustrate this statement.

The mongoose *(Herpestes auropunctatus auropunctatus)* is also known by Cuban farmers as the ferret, introduced in Cuba during World War I to control rats in sugarcane plantations. It is reported among the "10 most harmful invasive alien species in the world". It has a marked diurnal activity, generally observed active and in colonies between 10:00 a.m. and 4:00 p.m., especially during the rainy season. They usually feed on a variety of small animals, young vertebrates larger than themselves such as sea turtles, snakes and may also feed on fruits and vegetation if conditions do not offer any other alternative.

It is one of the main wild reservoirs of rabies in Cuba. The saliva of sick animals infected with rabies is the main vehicle of transmission, introduced into the host through bites or scratches. It causes great disasters by affecting sugar cane and corn plantations. It is responsible for the disappearance of lizard species.

The Perro jíbaro (*Canis lupus familiares)* is extremely dangerous for farmyard animals, as well as for wildlife, mainly vertebrates, where the almiquí, hutias, etc. stand out. The jíbaros dogs have been successfully established and are unlikely to be eliminated.

The jibaro cat *(Felis silvestres catus)* causes great havoc to both domestic animals and wildlife in natural ecosystems, mainly vertebrates.

Wild boar (*Sus scrofa)* is a domestic animal that was released or escaped. Introduced in many parts of the world, it causes damage to crops, reserves and properties causing many diseases such as Leptospirosis and Foot and Mouth Disease. By browsing, they uproot large areas of native vegetation and spread weeds, disturbing the development of ecological processes such as species composition and succession. They are omnivores and their diet may include juvenile tortoises, birds and endemic reptiles.

Others, such as the black rat and the brown or gray rat, are harmful species because they are carriers of ectoparasites in the case of the former and because they live in human settlements in the case of the latter, which is a pest not only because it devours food but especially because it transmits serious diseases.

The control and management of all these Invasive Alien Species depends to a great extent on the will of humans and the hygienic-sanitary measures that can be applied.

In this chapter you have studied a kingdom of great diversity and that shows important evolutionary aspects and adaptations to different forms of life, which is why it is considered, together with the plant kingdom, a kingdom of great zoological importance.

Self-assessment activities.

- Why, if the species that make up these groups are so different, do they all belong to the same kingdom?
- Do the sponges have tissues? Argument.
- What were the first animals to present tissues?
- Why are species belonging to the phylum Platyhelminthes triploblastic?
- You consider it important to study the life of nematodes. Argument.
- What is the digestive system of mollusks like?
- We can affirm that annelids are very small animals. Argument.
- Why do you think that arthropods comprise the largest group of species in the living world?
- Mention any echinoderm that causes harm to man. Why?
- Why can we say that fish, amphibians, reptiles, birds and mammals are chordates?
- What is the difference between the shark skeleton and the creole snapper?
- Why is it claimed that amphibians have double lives?
- Why can it be said that amphibians invade the land and

reptiles conquer it?

- Argue the importance of reptiles and birds in nature.
- Why are mammals the most evolved group of the Animal kingdom?
- Briefly explain how measures for the care and conservation of fauna are applied in Cuba.

CHAPTER 3: THE RELATIONSHIP OF ORGANISMS WITH THE ENVIRONMENT.

After having concluded the study of the different kingdoms and their representative specimens, it is necessary to know how each one of them establishes its relationships with the environment and among themselves.

In biology, the branch that deals with this study is Ecology. It studies nature as a great whole in which physical conditions and living beings interact with each other, establishing complex relationships.

Sometimes the ecological study focuses on a very local and specific field of work, but in other cases it is interested in very general questions. An ecologist may be studying how light and temperature conditions affect banana plantations at a certain time of planting, while another studies how energy flows in the rainforests of our country; but the specific aspect of ecology is that it always studies the relationships between organisms and between organisms and the non-living environment, i.e., the study of ecosystems.

The ecosystem is the level of organization of nature that is of interest to ecology. Observe the relationships established between the different components shown in the illustration of an ecosystem.

The ecosystem concept is particularly interesting for understanding the functioning of nature and a multitude of environmental issues.

We must insist that human life develops in close relationship with nature and that its functioning affects us completely. It is a mistake to consider that our technological advances: automobiles, big houses, industries, computerization, knowledge of the cosmos, etc., allow us to live apart from the rest of the biosphere, and the study of ecosystems, their structure and functioning, shows us the depth of these relationships.

Ecosystems are complex systems such as forests, rivers or dams, formed by a network of **physical elements** (the biotope) and **biological elements** (the biocenosis or community of organisms).

The ecosystem is the level of organization of nature that is of interest to ecology. In nature, atoms are organized into molecules and molecules into cells. Cells form tissues and these organs are assembled into systems, such as the digestive or circulatory systems. A living organism is made up of several anatomical-physiological systems intimately linked to each other.

It is the organization of nature at levels higher than that of organisms that is of interest to ecology. Organisms live in populations that are structured in communities. The concept of ecosystem is even broader than that of community because an ecosystem includes, in addition to the community, the non-living environment, with all the characteristics of climate, temperature, chemical substances present, geological conditions, etc. The ecosystem studies the relationships between the living beings that make up the community, but also the relationships with non-living factors.

The functioning of all ecosystems is similar. They all need a source of energy which, flowing through the various components of the ecosystem, sustains life and mobilizes water, minerals and other physical components of the ecosystem. The first and foremost source of energy is the Sun.

In all ecosystems there is also a continuous movement of materials.

The different chemical elements pass from the soil, water or air to the organisms and from some living beings to others, until they return, closing the cycle, to the soil or water or air, so that matter is recycled - in a closed cycle - and energy flows generating organization in the system.

When studying ecosystems, the knowledge of the relationships between the elements is of more interest than what these elements are like. Specific living beings are of interest to the ecologist because of the function they fulfill in the ecosystem, not in themselves as they may be of interest to the zoologist or botanist. For the study of the ecosystem it makes no difference, in a way, whether the predator is a ferret or a shark. The role they play in the flow of energy and in the cycling of materials is similar and is what is of interest in ecology.

As a complex system, any variation in one component of the system will have repercussions on all the other components. This is why the relationships that are established are so important.

Ecosystems are studied by analyzing food relationships, material cycles and energy flows.

a) Food relations.

Life needs a continuous supply of energy that reaches the Earth from the Sun and passes from one organism to another through the food **chain** (also called food chains).

Food webs (meeting of all trophic chains) begin in plants (producers) that capture light energy with their photosynthetic activity and convert it into chemical energy stored in organic molecules. Plants are devoured by other living beings that form the trophic level of **primary consumers** (herbivores).

The shortest food chain would be formed by the two links mentioned above (example: a cow feeding on vegetation). But herbivores are usually preyed upon by carnivores (predators) that are **secondary consumers** in the ecosystem. Examples of three-linked food chains would be:

grass - cow - man

Sometimes long chains are observed, for example:

grass - cricket - lizard - bird - man - grass - cricket - lizard - bird - bird - man

But food chains do not end with the top predator, but as all living things die, there are necrophages, such as some fungi or bacteria that feed on dead waste and detritus in general (**decomposer** organisms). In this way, the problem of waste is solved in nature.

Detritus (organic remains of living beings) is often the beginning of new trophic chains. For example, deep-sea animals feed on the detritus that descends from the surface.

The different food chains are not isolated in the ecosystem but form an interweaving and are often referred to as a food web.

A very useful representation to study all this trophic web are the **pyramids** of biomass, energy or number of individuals. In these pyramids, several levels are placed with their width or surface area proportional to the magnitude represented. On the lowest level are the producers; above them are the first order consumers (herbivores), then the second order consumers (carnivores) and so on.

b) Cycles of matter.

The chemical elements that make up living beings (**oxygen**, **carbon**, **hydrogen**, **nitrogen**, **sulfur** and **phosphorus**, etc.) are passed from one trophic level to another. Plants take them from the soil or the atmosphere and convert them into organic molecules (carbohydrates, lipids, proteins and nucleic acids). Animals take them from plants or other animals. They then return them to the soil, atmosphere or water through respiration, feces or the decomposition of corpses when they die. In this way we find in every ecosystem **cycles** of oxygen, carbon, hydrogen, nitrogen, etc. whose study is essential to know how they work.

c) Energy flow

The ecosystem is kept functioning by the **flow of energy** from one level to the next. Energy flows through the food chain in only one direction: it always goes from the sun, through the producers to the decomposers. Energy enters the ecosystem in the form of light energy and leaves in the form of heat energy that can no longer be reused to keep another ecosystem functioning. This is why an energy cycle similar to that of chemical elements is not possible.

It has already been stated that ecosystems are not isolated, but that the planet is made up of many ecosystems that also establish relationships among themselves, forming the integrity of all terrestrial spheres and of planet Earth itself.

A **biome** is a global classification of similar areas, including many ecosystems, climatically and geographically similar, i.e., an ecologically defined area in which similar climatic conditions and similar communities of plants, animals, and soil organisms occur, often referred to as large ecosystems. Biomes are defined based on factors such as plant structures (trees, shrubs, and grasses), leaf types (broadleaf and needle plants), distance between plants (forest, jungle, savanna), and climate. Unlike ecozones, biomes are not defined by genetics, taxonomy, or historical similarities and are often identified with special patterns of ecological succession and climax vegetation.

The simplest classification of biomes is:

1. Terrestrial biomes.
2. Freshwater biomes.
3. Marine biomes.

By observing the illustration below, you will be able to understand everything that has been analyzed so far, and you will see that food relationships, cycles and energy flow are closely related, forming a unit that constitutes ecosystems.

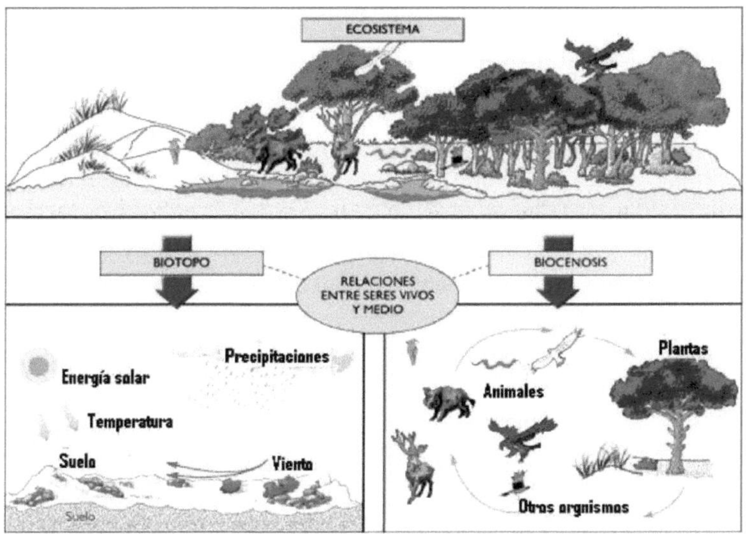

Eight major ecosystems (or **biomes**) are recognized on the planet. These are the temperate forest, tropical rainforest, desert, grassland, tundra, taiga, chaparral and ocean. Each is very different from the others because the amounts of sunlight, hence temperature and rainfall are very different. Likewise, each has special plants and animals that live there adapted to those conditions.

The boundaries of a biome are determined by climate more than any other factor. For example, because the northernmost biome, the tundra, is colder and has shorter growing seasons than the other warmer biomes, the tundra has fewer types of vegetation because few plants can tolerate the extreme low temperature conditions that prevail there. Tropical and subtropical biomes, on the other hand, which are found at lower latitudes, i.e. closer to the equator, experience a relatively smaller temperature range over the course of the year, more rainfall and a greater diversity of plants and animals develop there.

Next we will detail the characteristics of the different biomes and within them those that will be useful for the work of some subjects in primary schools, known as vegetation zones and animal population.

Tundra is located in the coldest latitudes, wherever snow melts seasonally. It is exposed to long, harsh winters and very short summers. In many places the sun does not set at all for many days in mid-summer, yet the amount of light at midnight is only one-tenth of that at midday. Low temperatures cause little rainfall and most of it occurs in the summer months.

Tundra soils are poor in nutrients and have few traces of organic material. All these characteristics, together with the fact that one layer of tundra soil always remains frozen and that its thickness and depth vary, interfere with drainage and prevent the development of large plant roots. All this presents to the eye a swampy landscape of extensive shallow lakes and slow-moving streams.

In view of the above, it is understandable that there are few species of organisms, although sometimes there are lots of moose, lichens (the misnamed reindeer moss), grasses. There are no easily recognizable trees and shrubs. Dwarf willows and other dwarf trees are common, rarely reaching more than 30 cm. in height.

Hardy permanent tundra animals include weasels, arctic foxes, snow hares, hawks, and others. In the summer, large herbivores such as musk ox and caribou come to the northernmost parts of the tundra to graze. There are no reptiles or amphibians, only a few insects in the summer.

Oil exploitation and military use have caused long-lasting damage to large portions of the Arctic tundra, which is likely to persist for many years.

The **taiga** or boreal forest (meaning the same as northern), which extends across North America and Eurasia, covers approximately 11% of the planet's emerged lands. Winters are extremely cold, although not as harsh as in the tundra. The growing season is also low, it receives little precipitation, its soil is acidic, poor in minerals, characterized by a deep layer of fragments of pine and fir partially decomposed on its surface. It has numerous ponds and lakes in places where depressions were left on the earth's surface as a result of previous glaciation.

Its vegetation is characterized by trees that lose their leaves in autumn such as poplar and birch, generally conifers that offer elegance to the landscape.

The fauna of the boreal forest includes some species such as caribou that migrate from the tundra to the taiga to spend the winter, wolves, bears and moose. There are also rodents and predators with fine fur such as squirrels, rabbits, lynxes, stoats, among others. Most of the bird species that live there are stationary and abundant but migrate to warmer climates during the winter. Insects abound, but amphibians and reptiles remain scarce as in the tundra, except in the southern extensions.

The soils of the taiga are not very productive, so very little agriculture is developed, however the boreal forest generates large quantities of wood, as well as skins from the animals that live there and other forest products, such as resins, etc.

Temperate forests are located in those latitudes where precipitation is relatively high and varies with latitude. Areas far from the coasts tend to be dry because areas of permanent high pressure prevent the accumulation of humid air, for example, the Sahara desert. Air passing over a large expanse of land can dry out without being able to recharge with moisture. However, the climate of the North American subcontinent is dominated by rainfall projected by mountain ranges, especially in the west.

Temperate coniferous rainforests can be found in southeastern Australia and southern South America. In this biome, precipitation is aided by condensation of water from dense coastal waters. The proximity of a temperate rainforest to the coastline moderates the temperature, so that seasonal fluctuation is reduced and consequently winters are mild and summers are cool. The temperate rainforest is characterized by relatively poor soils in nutrients, although their organic content can be high.

The dominant vegetation consists of evergreen trees, i.e. they never lose their leaves, such as firs, pinabetes and thujas. There are plants that grow on large trees without being parasites (epiphytes) such as mosses, lycopods, lichens and ferns.

Among the animals that live in these forests are squirrels, deer and numerous species of birds.

Temperate rainforests are the best producers of wood and pulp for paper in the world. It is also one of the most complex ecosystems on the planet, so indiscriminate logging would affect them considerably.

The **temperate deciduous forests** are characteristic of places where precipitation ranges between 75 to 125 cm. per year approximately. It has warm summers and harsh winters. Its soil is formed by an upper layer, rich in organic matter, and a deep lower layer, rich in clay.

These forests are dominated by hardwood and broad-leaved trees such as oak, walnut and beech, which lose their foliage annually. In the southern areas the trees are broad-leaved evergreens such as magnolia.

Throughout the world, deciduous forests were the first to be transformed for agricultural use. For these reasons many species are now extinct such as pumas, wolves, deer, bison, bears and other species of small mammals and birds. There were also reptiles and amphibians, as well as a wide variety of insects.

Where these forests have been allowed to regenerate due to better resource management, they are in a semi-natural state, i.e., heavily modified by man.

Temperate grasslands are found in temperate zones with moderate rainfall where summers are hot, winters are cold and rainfall is often scarce. Their soil is rich in organic matter as the aerial parts of many grasses die each summer and contribute to the organic content, while roots and rhizomes survive below the surface. The Midwestern United States is an excellent example.

temperate grassland. Few trees grow, except near rivers and streams, but grasses abound. In other times, it was characteristic to find tall grass species, predators such as wolves and coyotes. Prairie dogs, foxes, black-footed ferrets and birds of prey, prairie birds, lizards, snakes and a large number of insects were also found. The current lack of many of these species has been determined by man's misuse of these biomes.

Steppes, or short grass prairies, are temperate grassland habitats in which there is less precipitation than in wetter grasslands but more than in deserts, they have less grasses and the soil is bare. Native grasses are drought resistant. They present the ideal conditions for cultivation as cereals, which is why in the places where they exist, they rarely maintain their original conditions, as they have been used in agriculture.

The **charrapales** are habitat of mild winters combined with very dry summers. Their soil is thin and not very fertile. Fires are very frequent in them, particularly in late summer and autumn.

Charrapal vegetation is remarkably similar in different regions of the world, although the individual species are very different. Evergreen shrub species abound and may contain drought-tolerant pines and oaks. Trees and shrubs often have tough, small leaves surrounded by a thick cuticle that reduces water loss.

Some animals, such as reptiles, have their bodies covered with structures resistant to the action of strong temperatures and their coloration is similar to that of the place where they live in order to attenuate the effect of the scarce vegetation.

Deserts are very dry areas located in temperate and tropical regions. The low water content of the atmosphere causes extreme temperatures. Some deserts are so dry that they are devoid of plant life, as for example in the deserts of Namibia and the Atacama-Sechura in Chile and Peru. For such reasons the soil is poor in organic matter but often high in minerals.

Vegetation cover in the deserts is scarce or non-existent as already mentioned. Where it exists it is represented by cacti and various grasses growing in scattered clumps. Many desert plants are endowed with defensive spines to resist intense grazing pressure.

Desert animals are small. During the heat of the day they remain under cover or return to their shelter periodically. At night they come out to graze or hunt. There is a diversity of insects adapted to desert life, as well as reptiles (lizards, tortoises and snakes). Mammals include rodents such as the American kangaroo rat, which does not need to drink water; it uses water from its food or that generated by its body during metabolism. Hares, kangaroos, rabbits and some birds of prey live.

American deserts have been disturbed by man, as vehicles damage the vegetation; some cacti and turtles have almost disappeared due to indiscriminate collection and hunting.

Savanna is a tropical grassland biome with widely dispersed clumps of low trees. It is found in areas of low rainfall, with prolonged dry periods and the annual temperature variation is small. For this reason the seasons are regulated by precipitation rather than temperature.

Savanna soil is low in mineral nutrients because the parent rock from which it was formed is poor.

It is characterized by wide expanses of grassland interrupted by occasional trees such as acacias, whose barbs protect it from herbivores.

The African savanna is home to the largest assemblage of hoofed mammals in the modern world: large herds of herbivores such as blue wildebeest, antelopes, giraffes and zebras. Large predators, such as lions and hyenas, kill members of these herds. These herds often migrate in places with seasonal rains to other parts.

Tropical rainforests are characterized by the heavy rains that occur in them where the water that falls is practically from the same location that is evaporated by the transpiration of the trees that live there.

Because the temperature is high year-round, decomposing microorganisms, ants and termites rapidly disintegrate organic matter and its nutrients are absorbed by highly developed mycorrhizae and then transferred to plant roots. In this way, the nutrients of tropical rainforests are trapped in the vegetation, not in the soil itself.

In these forests, productivity is very high due to the significant solar energy input and abundant rainfall, and many photosynthetic processes take place, so oxygen production is high.

In the forests there is a distribution of vegetation in tiers, the highest being occupied by the treetops, the middle one by shrubs that practically do not allow the passage of light to the lower level represented by smaller plants specialized for life in these conditions.

The trees of these tropical rainforests are evergreen with flowers, their roots form a carpet of almost 1 meter thick on the soil surface, which captures and absorbs almost all the mineral nutrients released from the leaves and detritus by decomposition processes. These trees support a wide diversity of epiphytic plants.

The animals of this tropical rainforest include insects, reptiles and amphibians more abundant and varied than in any other biome. Birds are varied and often brightly scented. Mammals such as sloths and monkeys live in the trees, although some large mammals such as elephants are ground dwellers.

Unjust and ambitious human action and industrialization in tropical countries may mean the end of most or all tropical rainforests. It is thought that many rainforest organisms will become extinct in this way before they have been scientifically described.

There are other ecosystems that are not terrestrial as described above. Rivers and streams are flowing water ecosystems, while lakes and ponds are stationary water ecosystems. There are also marine habitats. It is not considered necessary to elaborate on them in this chapter.

From the human point of view, many see ecosystems as production units similar to those that produce goods and services. Among the most common goods produced by ecosystems are timber and fodder for livestock. Bushmeat can be very profitable under a well-controlled management system as is the case in some places in South Africa and Kenya.

Ecosystem services derived from ecosystems include: **Enjoyment of nature**: which provides sources of income and employment in the tourism sector, often referred to as ecotourism. **Water retention**: facilitating better water distribution. **Soil protection**: an outdoor laboratory for scientific research.

A greater number of species or biological diversity **(biodiversity)** in an ecosystem makes it more resilient because a greater number of species can absorb and reduce the effects of environmental changes. This reduces the impact of environmental change on the overall structure of the ecosystem and reduces the chances of a change to a different state. This is not universal; there is no proven relationship between species diversity and the capacity of an ecosystem to provide goods and services in a sustainable manner. Tropical rainforests produce very few direct goods and services and are highly vulnerable to change. In contrast, temperate forests regenerate rapidly and return to their former state of development within one human generation, as can be seen after forest fires.

Whatever the biome we are talking about, it requires the necessary attention for its care and conservation. Today, the world is witnessing events that considerably affect natural resources and the organisms that live in them. Indiscriminate logging, hunting of endangered species and, in other cases, abuse of hunting for commercial purposes, as well as the use of resources for fuel, food, etc. are some of the ways that cause great damage to the environment. The environmental policy in the world and in Cuba makes that these affectations are reduced and those resources that were affected can be reestablished. However, this requires a greater effort.

Cuba has created protected areas, national parks, biosphere reserves and other projects that guarantee better control and management of the different species and resources that must be guaranteed for a sustainable future.

In this chapter, we have presented the relationships that organisms have with their environment and the different ways in which they adapt to the conditions in which they live. The work carried out for the conservation and protection of all ecosystems and, consequently, of the environment in general, is considered of great importance.

Activities for self-monitoring

1. - Briefly state what Ecology studies.

2. -After reading the theoretical conceptions regarding ecosystems, group the common elements and define for yourself what an ecosystem is.

3. - Draw a pyramid of ecosystems and place on it the position of each of its components.

4. - Mention the main components of a food chain. With this in mind, construct a short and a long food chain.

5. - What are the main climatic and soil factors that are considered to classify terrestrial biomes?

6. - Make a table to compare the different terrestrial biomes studied in terms of climate, soil characteristics and representative organisms.

7. - Which biomes are most favorable for agriculture? Justify your answer.

8. - Explain briefly how human attitudes and the development of industrialization have affected the stability of the terrestrial biomes studied.

9. - Consult different bibliographic sources and summarize what measures are taken for the care and conservation of ecosystems and other natural resources in Cuba.

BIBLIOGRAPHY .

1. Abreu Alfonso, O. (1990) La educación ambiental: una acción de todos. In *Revista Técnica Popular,* Havana.

2. Armiñada García R. (2020) *Nociones de Zoología de los no cordados.* Digitized Material.

3. Arredondo Antúnez, C. et al. (1996) *Zoology of chordates.* Part One. Editorial Pueblo y Educación. Ciudad Habana.

4. Arredondo Antúnez, C, et al. (1996) *Zoology of chordates.* Second part. Editorial Pueblo y Educación. Ciudad Habana.

5. Balmaseda Meneses M. J et al. *Biología General.* Digital text.

6. Barnes D. R. (1986) *Zoología de los invertebrados.* First Part, Revolutionary Edition, Havana City.

7. Barnes D. R. (1986) *Zoología de los invertebrados.* Second Part, Revolutionary Edition, Havana City.

8. Bauzá Aguiar, Vivian et al. (1984) *Curso Facultativo de Biología*. Editorial Pueblo y Educación. La Habana.

9. Berovides Álvarez, V. (1985) *Ecología, ciencia para todos.* Editorial Científico- Técnica, Havana.

10. Clarke, G.L.(1971) *Elementos de ecología.* Editorial Revolucionaria. Havana.

11. Cuevas, J. R. and F. García (1992) *Los recursos naturales y su conservación.* Editorial Pueblo y Educación, Havana.

12. University for all (2007) *Course on Biological Diversity.* University for all. Editorial Academia. Havana.

13. De Robertis, E. D. P., W. W. Nowinski and F.A. Sáez (1972) *Biología Celular*, 8th Edition, Editorial Ateneo. Buenos Aires.

14. De la Torre Callejas, S. (1987) Zoología *de los invertebrados inferiores.* Editorial Pueblo y Educación. Ciudad Habana.

15. Kourí Flores, Juan B. (1978) *Biología General 1,* 10th grade. Editorial Pueblo y Educación. La Habana.

16. Lau Apó, Francisco et al. (2004) La Enseñanza de las Ciencias Naturales en la escuela primaria. Editorial Pueblo y Educación. La Habana.

17. Ministry of Education and Culture of Spain (1986) Curso de formación de profesores de Ciencias. Volumes: Energy; Ecology; Living beings. Iberoamerican Television, Madrid.

18. Multisaber (2003) Software: *Mysteries of nature.* Havana.

19. Valentín Arbona, Marta (1988) *Botánica Sistemática I.* Editorial Pueblo y Educación, Havana City.

20. Varona, Luis S. (2005) Mammals of Cuba. Editorial Gente Nueva. Havana.

21. Vilee,C. (1996) Biología, 3rd edition, Editorial Iberoamericana, Mexico City.

22. Zamora Martín, E. (1980) *Diccionario de términos biológicos.* Editorial Científico-Técnica. La Habana.

23. Zilberstein Toruncha, José, et al. (1991) Biology 5, 12th grade. Grade. Part 2. Editorial Pueblo y Educación. La Habana.

Printed by Books on Demand GmbH, Norderstedt / Germany